journey for one

jodie hopkins

journey for one

A Guide to Gaining the Courage and Skills to Travel Solo

Copyright © 2020 Jodie Hopkins

All rights reserved. No part of this book may be reproduced in any form or by any electronic or mechanical means, including information storage and retrieval systems, without written permission from the author, except in the case of a reviewer, who may quote brief passages embodied in critical articles or in a review.

Trademarked names appear throughout this book. Rather than use a trademark symbol with every occurrence of a trademarked name, names are used in an editorial fashion, with no intention of infringement of the respective owner's trademark.

The information contained in this book is for general information and entertainment purposes only. The recommendations, opinions, experiences, observations, or other information contained herein is provided "as is" and neither the author nor publisher make any representations or warranties of any kind, express or implied, about the accuracy, suitability, reliability or completeness of this book's content. Any reliance a reader places on such information is therefore strictly at their own risk. All recommendations are made without guarantee on the part of the author and publisher. To the maximum extent permitted by law, the author and publisher disclaim all liability from this publication's use. In no event will either author or publisher be liable to any reader for any loss or damage whatsoever arising from the use of the information contained in this book. This book is not a substitute for professional services and readers are advised to seek professional aid in the event of emergency.

Printed in the United States of America

First Printing, 2020

ISBN 978-1-7329022-2-0

Positively Powered Publications
PO Box 270098
Louisville, CO 80027
PositivelyPoweredPublications.com

Cover design by Melody Christian.
Interior layout by YellowStudios.

DEDICATION

To my incredible mother, JoAn Hopkins, who is the strongest, most generous, and loving human being I know. Thank you for teaching me that even a trip to the laundromat can be an adventure!

CONTENTS

Introduction 1

Chapter 1   Why Travel Alone? 11

Chapter 2   Making the Decision to Go Solo 19

Chapter 3   Knowing Yourself 31

Chapter 4   The First Step 41

Chapter 5   Setting Your Goals 49

Chapter 6   The Joy of Success 61

Chapter 7   When Things Go Wrong 69

Chapter 8   Staying Safe 85

Chapter 9   Health Matters 97

Chapter 10   Communicating with Home 113

Chapter 11   Explaining Your Plan to Others 121

Chapter 12   Making a Plan 131

Chapter 13   Being Flexible 143

Chapter 14   Fear 155

Chapter 15   Ready, Set, Go 171

About the Author 181

Acknowledgments 183

Resources 185

INTRODUCTION

I've learned that fear limits you and your vision. It serves as blinders to what may be just a few steps down the road for you. The journey is valuable, but believing in your talents, and your self-worth can empower you to walk down an even brighter path. Transforming fear into freedom—how great is that?

—Soledad O'Brien

Welcome to Your Adventure

Welcome aboard. I am Jodie Hopkins, your guide for this amazing journey. I am a woman who loves to travel the world alone and with others. I have visited 33 countries, 6 continents, and all 50 of the United States. I have taught students from over 100 countries how to speak English, and I want you to learn to love traveling alone as much as I do. I am incredibly happy that you are with me for this important adventure.

I have been traveling my whole life. Several family vacations a year was normal in our house and we often traveled to see new things and experience places that were different from what we were used to. I have been traveling alone for a little over 40 years. Sometimes that solo travel was for business, most of it for fun. Since I was a teacher, I had a fabulous vacation schedule and took advantage of that time to travel as often as possible. I have had marvelous trips and I've had a few that were less than enjoyable. That's the nature of the beast. I have traveled with family, friends, students, and colleagues. I really like being independent. I like the freedom it gives me and how it lets me avoid the inevitable challenges that come with coordinating travel plans with others. Over the years I have come to believe that, for me, traveling on my own offers unique and valuable opportunities that I cherish. Some of my most tremendous insights and greatest growth have come from those times in my life when I have ventured out on my own and found comfort, joy, and insight from my own company.

I was lucky enough to be born into a family that loved to travel and truly enjoyed experiences in other countries and cultures. Meeting interesting people, having cultural experiences, and exploring foreign lands was just what we did. It was how my parents raised my two sisters and me. How lucky was that? Ever since then I have traveled every chance I could get and literally have scrapbooks filled with photos, ticket stubs, and memorabilia from my travels.

The first time I remember traveling alone was running away from home at age 5. When I ran out of cookies from my bandana tied to a stick, I came home. After that, my true first solo adventure was a trip to Mexico at age 14 with a school group to study Spanish. I am still astounded that I made that trip. I was a *very* shy child and was scared of my own shadow, but somehow, my parents and I trusted my teacher, Mr. Hanlon, and off I went. That was the first of my many amazing solo trips. In the time since then I have traveled to 33 countries of the world, visited six of the seven continents, and have lived on my own in Vienna, Austria for 11 years.

Those travels, many of which I did alone, have been the highlights of my life and have given me joy I never thought possible, which brings me to why I wrote this book. I started noticing that many people, mostly women, were telling me they could never travel alone like I do. They said they were just too scared and didn't know how to do everything needed for traveling alone. I started to think about how I got started and how I overcame the fear and lack of skills that I had when I began. Over time, I thought that my stories and my teaching experience might be able to help others enjoy the adventures and sights as much as I did. That's when the book, *Journey for One*, began. I truly hope that it gives you the courage and the skills you need to overcome your fear and give solo travel a try. It is also one of my greatest hopes that you will find this book a fun and engaging way to learn about and practice those skills.

My time alone when I travel is some of the most valuable time I ever have. It's a time to get to know myself: to reflect, dream, figure out, and find peace for my life. That, for me, is one of the greatest gifts of solo travel. There are not many times in life when you are somewhere beautiful and have the uninterrupted time to really immerse yourself in your own thoughts. One such trip for me was a trip to Portugal I took a couple of years ago. I was living in Vienna, teaching at the American school there, and was strongly considering retirement. When I tried to think about it at home I was always distracted by dishes to be done, emails coming in, meetings at work, or activities with friends. It wasn't until I was able to go to Portugal, and sit by the sea uninterrupted for a few days, that I was able to have the time to figure out the pros and cons, the hopes and fears, and the joys and sorrows of retiring and moving back to the United States. I am so grateful I took that time alone to have those thoughts, for it brought me to a very good decision and put me in a situation to create a wonderful new life for myself. I could never have had that outcome, I believe, if I had not had the opportunity to spend several days alone.

The logical question then is, "Do you ever get lonely when you travel alone?" The simple and honest answer is yes. There have been trips when I wished I had been with someone else there to share the moment. My trip to India was like that for me. I had wanted to visit India forever, but had some fear about going there alone. I made the decision to go and experience it on my own anyway. After I got there and started traveling around I became lonely for someone that I could have shared the amazing events of that trip with. To say to someone, "Oh, my gosh! Look at that! There are seven people on that one motorcycle!" or "I am so sad about the dreadful conditions in the ghetto!" My experience in India was amazing and I can't wait to go back again, but everything there, to me, was brighter, bigger, harsher, more colorful, and more exotic than anywhere else I had ever been. It would have been great to share my amazement with a friend. So, you see, being alone may or may not be a lonely experience.

If life has taught me anything it is that there truly is no "one size fits all" in life. We all have different backgrounds, needs, personalities, and preferences. What may be thrilling to one is torture to another. It's kind of like having to sit through a science fiction movie when you *truly* only enjoy comedies. It doesn't make science fiction wrong, it's just not your thing. Travel is the same way. But... here's the catch. You never know if you're going to like science fiction if you don't ever go see a science fiction movie. You can't say you don't like sushi if you've never tried it. Solo travel is no different. Until you have tried it, you have no idea if you will really like it. I had a professor in graduate school who said that you couldn't judge whether or not you liked or could benefit from something until you had tried it three times. I would suggest to you that learning to travel alone requires the same effort. Make a first trip near home so you can get the feeling of doing it but always have an easy way out if need be. The next time, go somewhere that's a bit more of a distance. Maybe it requires a plane ride or a lengthy drive. Make it a little less familiar and a bit more of a risk. After that, when all goes well, the third trip could be to somewhere even more special and a place that really tests your solo adventurer skills. Think of it as exams in a class. The first adventure is the pre-test, the second is your midterm, and the big one at the end is your final. By then you will know for sure whether solo travel is something you will enjoy.

I want to be clear at this point that just as not everyone loves science fiction or tuna fish or snowshoeing, neither is everyone going to love solo travel. That's perfectly OK. You will get no challenges or judgment from me. I am not here to *make* you do anything. I am assuming that because you picked up this book, you have some interest or need to travel on your own. I will also assume that by reading this introduction you are at least a bit committed to trying to make it work for you. Therefore, please give yourself the gift of time and be patient with yourself as you give this book a try. Work your way through the process and wait until the end to decide if solo travel is for you. If nothing else, you will get the chance to do some great self-reflection and have some fun small adventures. It's all for you.

How to Use *Journey for One*

The first and most important thing you need to know about this book is that it is for you... only you. It is your own first solo travel journey. You get to go through the process all on your own. There are no right or wrong answers,

♦ Introduction ♦

no judgment, and certainly no passing or failing. In the end, your journey will be successful. I assure you.

The next thing you need to understand is that while this is your journey and only you can decide how it goes, you are not alone. I will be with you every step of the way. As a teacher for many years, I have set this book up in the same way I always did for my students. It is designed for success. Each chapter gives you an opportunity to reflect on your work, have a final task to complete, and a celebration of your accomplishments. So, put aside any old worries about learning that may still haunt your mind and enjoy the ride.

Each chapter is about a topic related to solo travel and walks you through a step-by-step process to learn and practice the skill(s) related to that topic. Each chapter includes the following sections:

- **QUOTE**: This quote gives you a mindset to hold onto as you proceed. It's a little bit like the life ring on a ship. It is something you can hold onto so that you feel secure.
- **TRAVEL TALE**: These are stories about real life adventures related to the topic for that chapter. The stories are often from my own travels, but not always.
- **PACK YOUR BAG**: This section allows you a chance to give some thought to what you already know and can take with you about the topic before you start thinking about new information. It sets you up to be ready for learning.
- **LEARNING THE WAY**: As the name indicates, this part of the chapter is where I introduce what may be new information, food for thought, tips and tricks, and ideas that help you see the topic in, perhaps, a new light.
- **ADVENTURE**: I can't even think of trying to teach you important information without some good, old fashioned practice. The Adventure section will give you a chance to try out what you learned in the Learning section. It is an important part of the process because as many wise educators know, "Tell me and I forget, teach me and I may remember, involve me and I learn." Please do the adventures. I know it's tempting to sit on the couch with your book and be happy just reading, but the adventures are part of building a foundation for your courage and

skills to grow. If you don't do them, you will only be short-changing yourself.

- **LOOKING IN THE REARVIEW MIRROR:** Just as the name indicates, you will be looking back at what you did in the Adventure section and reflecting on how things went. What did you learn? What worked well? What didn't go so well? What would you do differently in order to be a more effective solo traveler next time?
- **CELEBRATION:** This is a very important section of each chapter. I urge you not to skip this part, because it may take a little time or effort on your part. It is an important element to learning to travel on your own. Everything I ask you to do in the book has a specific purpose in building your confidence and skill. To cut out any activity or part of the process in this book is like playing adult hooky. You miss out. You could end up missing a very important skill that will make the difference between success and defeat when it's time to travel in the real world. I truly urge you not to shortcut the process.

There are a couple of terms that I use in the book that I want to explain. They may be teacher jargon and I'd hate to have you be frustrated because you weren't sure what I meant.

Brainstorming: Many of you may have participated in this process in school or at work. It is an activity where you write down each and every idea or thought that you have about a topic. Don't worry whether or not it's good or bad, just write it down. You'll get a chance to prioritize and sort out the "weeds" later.

I have provided lots of space for you to write in each chapter where there are writing activities involved. Please don't ever feel that you are under an obligation to fill up the whole space. I'm guessing that as you go through the book, there will be times when you will use only a line or two and other times when you will need to write on additional pieces of paper. Either way is fine. Remember, it's for you so do what makes sense and meets your needs.

People sometimes confuse the concepts of alone and lonely. For me, being alone is not a bad thing. It means you are on your own, there is nobody else to tell you what to do, how to do things, or when things have to happen. You are free to choose activities and make decisions all on your own. You are,

• Introduction •

at the same time, free to and yet responsible to make your time be whatever you'd like it to be. It means you have no one present to talk to or to discuss things with and that any decisions you make are all up to you without the benefit of the input of another. Lonely, on the other hand, means that you feel alone even when your friends or loved ones are physically there. You somehow feel left out, overlooked, unappreciated, or out of the loop. There could be a thousand reasons for why you are feeling lonely, but the basic idea is that lonliness can happen whether or not others are present.

Based on those definitions, I can travel alone but not be lonely. I can be free to choose my destination and my itinerary, and enjoy them thoroughly because it's *my* vacation, *my* adventure. I can also enjoy the ability to not have to consult anyone else about where to go, what to do, or where to eat. I can be my own best companion and have the most wonderful time being alone with my thoughts and reflections.

When I started writing this book, one of the things that was tough for me was the idea that I wouldn't be there in person to get you started. It's like a coach not being able to give the pep talk at the beginning of a game. So, instead, I have decided to give you a little send-off in writing. I really want you to know and believe that you have everything you need to become a courageous and skilled solo traveler. You can do this. Please remember...

- You are the boss of your own journey. You are free to do or not do anything in this book. Just remember that, like baking a cake, if you don't follow the directions and put in the right ingredients, you're not likely to end up with a very good cake.
- It's important that you are completely honest with yourself as you go through this process. It's like going on a diet. You can sneak away and eat that forbidden candy bar and not tell anyone you cheated, but ultimately it is only you that is affected by the cheating. Be honest. No one is going to see or judge your notes unless you show them. You can only overcome the fear and learn the skills you need by honestly assessing your own thoughts and abilities. You have way more to gain than you have to lose.
- This workbook is designed to be followed from beginning to end in order. You may be tempted to bounce around by going first to the chapters that sound the most interesting or helpful to you. It's OK to

peek and see what lies ahead but I urge you to follow the order of the lessons and exercises to get the most from this journey.
- Be gentle and kind with yourself. Everyone is starting from a different place with a different set of experiences and needs. You are undertaking something that challenges you to think differently and to do things that may not yet feel comfortable. Take it one step at a time and give yourself the love and kindness that you deserve and would offer to anyone else trying to do something new and a little frightening.
- Feel free to keep your journey with this book to yourself. You know yourself best. If you find comfort in telling others what you are doing so that they can give you support and encouragement, great. If not, you are welcome to do this work and develop these skills in your own way, in your own time, and on your own. The important thing is to feel safe and free to do this work without criticism or judgment. I'll leave the decision completely up to you.
- There are no absolute right or wrong answers to this work. What works for one may not be as helpful for another. When you are filling out the exercises or doing the writing, don't worry about what I or anyone else might think of your responses. You are not being graded on this work and there are just as many ways to travel successfully as there are places to travel. If you feel like you messed something up or didn't like an exercise, good! That means that you tried it and thought about it. That's all I ask. Just give it a chance and use your own best judgment.
- In the end, it is you and you alone that will decide if you want to travel solo. You may go through this book and immediately get online and book a ticket to some wonderful destination, or you may decide that all of this thought and learning has helped you make the decision that solo travel is not for you. Either way, I am happy and I want you to be happy with your decision, too. We get chances in life to do many things. Sometimes we say no thank you to those challenges and chances and at other times we grab them, hold them tightly in our hands and shout *yes please*! Only you know what is right for you. My goal is to be sure that should you decide that you do want to travel on your own, that you have had the opportunity to be the most prepared and

most courageous you can be when you start planning for your first trip on your own.
- Last but not least. Have fun! Be ready to laugh at life, at your mistakes, and at yourself. I can't even begin to tell you the number of times over the years that I have either made a mistake, gotten things wrong, or started off unprepared and ended up shaking my head and laughing at myself for how things turned out. This is not a course in learning tax law or advanced physics. It is learning about yourself and that is something you already know a great deal about. When you can be your own best friend, you already have a travel partner for life.

Traveling solo can be rewarding, insightful, freeing, and life changing. At the same time, for some, it can be less than comfortable. I want you to know that no matter what you decide at the end of our journey together through this book, I hope that you find that this Journey for One was worthwhile, valuable, and a good experience for you. If you should decide that solo travel is not for you, that's OK too. You know yourself and what works best for you. Whatever you decide about traveling solo, I truly wish you bon voyage.

Now, it's time to put your seat back in the upright position, stow your tray tables, and fasten your seatbelt. We are about to take off on an incredible Journey for One. Please sit back, relax, and enjoy the adventure.

WHY TRAVEL ALONE?

Travelers are dreamers who make their desires for adventure a reality.

—Author Unknown

Travel Tale

I stood silently in my boss' office two weeks after I had started my new job. My boss chatted away about the conference he was "letting" me go to in St. Louis, Missouri. My ears had ceased hearing, my eyes were focused straight ahead, and my brain was spinning faster than a jet engine. Alone? I was going to this conference alone?

I had traveled a lot growing up. My parents had transported us often in planes, boats, trains, and our trusty motor home. I had visited Mexico, Canada, and almost every one of the 50 American states. But solo? I had *never* even briefly thought about traveling completely alone. What would I do? I wouldn't dare tell my new boss that I couldn't go. I couldn't possibly admit that the idea of going somewhere by myself scared the heck out of me. I knew I had to go and I would have to go alone.

Being a teacher of small children for many years and surviving the experience had led me to believe that, literally, I could handle anything. So, I sat myself down and made a list of the things that scared me about being by myself on a trip. The list looked something like this:

- Sleeping alone in a hotel.
- Eating alone at a restaurant.
- Sleeping alone in a hotel.
- Getting from the airport to the hotel and back.
- People would stare at me if I was walking around alone.
- Sleeping alone in a hotel.
- I would stand out by being alone.
- I would have no help if something went wrong.
- Sleeping alone in a ...

You get the idea.

I had about a month until the conference, so, just like I did for my young students, I planned a series of activities before then that would make me try out some of these experiences near home. That way, at least when I got to St. Louis it wouldn't be the first time I had ever even tried to do them. I pulled out my credit card, got on the telephone and began making reservations. I had one month to pull myself together.

On a drizzly Saturday morning two weeks later I grabbed my overnight bag, got in the car, and headed to a little town near me on a lake that I loved. It had good shopping and plenty to do to keep me busy. I pulled into the hotel parking lot. I must have sat there for about 15 minutes before I gathered the courage to go inside to check in.

The woman at the front desk took care of the usual paperwork and then asked, "Will it just be you staying tonight?" I knew it. I had already been identified as the pitiful lone traveler. I knew in my head that she was wondering what was wrong with me that I couldn't get anyone to come with me. I was on the verge of tears. She looked up, gave me the biggest smile ever, and said, "I'm so jealous. I wish I could go on a little holiday by myself." Suddenly, everything had changed. Rather than feeling like the freak of the travel world I was standing a little taller because I was doing something that another woman dreamed of doing. Now I was determined to make this work. I could do this.

I would be lying if I said I got a good night's sleep or that dinner alone at the restaurant that night was amazing, but I also know that the way things fell into place on that rainy Saturday in the Rocky Mountains gave me the motivation, courage, and strength to make the trip to St. Louis something I could do and even more importantly, something I could enjoy.

Pack Your Bag

It is now time to spend a few minutes alone with your thoughts. Think about why you are making the journey with this book. Why are you thinking of traveling alone? What do you hope to accomplish by traveling alone? How do you feel about the prospect of traveling alone? What are your thoughts about people who travel alone? On a scale of 1-10, ten being the highest, how excited are you about traveling alone?

After you have had some time to think, jot down those thoughts. It is important to write as much as you can. These thoughts will help at another point in the process, so the more you write now the easier it will be for you later. There is no specific length or style of what you write. It just needs to be meaningful to you.

_____

_____

_____

_____

_____

_____

_____

_____

_____

Learning the Way

I can think of as many reasons to travel alone as there are people who travel. It might be that you have recently found yourself living solo for the first time in a long time. Perhaps you have a partner who prefers to stay home. Maybe, like me, you are single and many of your friends have partners, children and/or responsibilities that don't allow them to get away. Whatever the reason ... *Welcome*! You are not alone! That sounds pretty strange when we are talking about traveling alone, but it's true. The number of solo travelers has increased dramatically over the last few years and more and more people are seeing the world on their own.

Why you are sitting here now with this book is not as important as the fact that you have taken a very important step in a wonderful journey. Getting more information, learning from others and trying things out is always important. Let's say that you decide to buy a new pair of dancing shoes. You would first consider what kind of shoes you need or want, think about which stores or brands might be best for your situation, and figure out how much you can afford to spend on the shoes. Then, you would probably ask others who have previous dance experience for advice or recommendations about these shoes. Only then would you make the trip to the stores to try on the different shoes to see which shoe actually fits you the best. Learning to travel solo is no different.

After thinking about the possibility of traveling alone, you started your quest for information. You picked up this book and perhaps looked online for information, checked travel guides or magazines, and then maybe talked to others who have traveled alone.

This book will not tell you where to go, what to see or how to get there. But, it will give you the opportunity to get information, learn from people who have "been there, done that," and try some activities out to see if solo travel fits your needs, personality, and lifestyle.

I would never expect someone to jump from never going to a fast food restaurant alone to traveling alone for six weeks in a third-world country in one step. It's a process: a journey; truly a Journey for One. No one can decide for you. This is a journey that requires you to know yourself, look at your likes and dislikes, and stretch yourself a bit as you venture down the path.

When you learn to dance, certain steps or moves come more easily than others. Some steps may leave you tired or a bit sore, others will give you even more excitement about moving on. Regardless, it is all part of the learning process and an important element of the journey. So, stick with it. Don't give up because you stumble once in a while. There are no time limits, no expectations, and no grades. This is about gently guiding yourself through unfamiliar territory. Before long it won't be unfamiliar and you will wonder why you didn't do it sooner.

With that said, I invite you to settle in and begin the step-by-step process of getting ready for your Journey for One. Have a wonderful journey.

Adventure

At a training I once attended, the trainer gave us 30 seconds to look for as many blue things as we could find in the room. The 30 seconds went by quickly, with people scurrying to try and remember all of the blue things they could see. After the time was up, he then asked, "OK. How many red things did you see?" What? He asked us to count the blue things. The message in the lesson was that we find what we are focused on looking for. It's the same idea as when you never see a certain model of car until you buy one and then suddenly everybody on the road drives one.

For this adventure, you are going to look for, notice, and remember the places that you see during your daily activities where you see people participating alone. Obviously, places like the gym or the grocery store will have significant numbers of individual participants. Pay attention at places like the mall, museum, library, sporting events, restaurants, concerts, movies, lectures, etc. Try to pay particular attention to activities that you think you would enjoy when traveling.

After you have observed solo participants in several locations, I would like you to choose an activity that you have not done alone before and enjoy it by yourself. As you attend this activity, focus not on yourself and being alone, but rather on who else is alone. What are they doing that might be helping them or could help you to be more comfortable being alone? Remember your thoughts and observations so you can record them afterwards in the Looking in the Rearview Mirror section below.

• Journey for One •

Looking in the Rearview Mirror

You have now had an opportunity to observe and engage in activities that people can do alone. Jot down your thoughts and observations from this adventure. What did you notice? What ideas did you get that might be helpful to you? What did you see that may have helped you with fears you were having about doing things alone? If you were to do the same activity alone again, what might you do differently next time?

_____
_____
_____
_____
_____
_____
_____
_____
_____
_____
_____

Celebration

Congratulations! You have made it through the first leg of your journey. You have taken your first steps toward being a happy and confident solo traveler. Now it's time to celebrate. Psychologists say that it is important to celebrate our successes for several reasons. One of those reasons is to recognize what worked and why it worked, so that you can learn and recreate that success. Another reason is to develop what they call a success mindset. It helps you to focus on the fact that you can and will be successful.

Celebrations are also helpful because they provide additional motivation for us to continue and to do well.[1] In light of that information, I would like you to think about something that you could do for yourself that would feel like a reward or celebration to you. Consider having that celebration be something that would also be possible in a traveling situation. If your idea of a celebration is to curl up with a good book under your grandmother's quilt, fabulous! Perhaps, on a trip, you could curl up by the pool or in a nearby park. The hope here is that you will begin making solo celebrations a regular part of your solo travel ritual. Below are a few ideas of travel-friendly celebrations.

- Sitting down with a favorite beverage.
- Watching a favorite or new movie.
- Working out at the gym.
- Going for a walk.
- Taking time to read.
- Looking back over photos that have been taken so far.
- Treating yourself to a nice dinner.
- Doing some form of your favorite hobby.
- Listening to music.
- Playing music.
- Singing.
- Doing artwork.
- Having a glass of wine.
- Buying yourself flowers.
- Other:_____

_____

OK. Find a celebration of your own that you can truly enjoy. You have made a good start. It's important for you to recognize your effort, strength, insight and commitment. When you travel solo it is your responsibility to congratulate yourself and recognize the growth that you have made. Now, enjoy the celebration of your progress on your Journey for One.

---

1   Brilliant Living HQ, "6 reasons why you should celebrate success," https://www.brilliantlivinghq.com/6-reasons-why-you-should-celebrate-success/

MAKING THE DECISION TO GO SOLO

You are in the driver's seat of your life and you can decide to keep driving down the road you are on, or you can turn and go a new direction whenever you want to.

—Jerry Bruckner

Travel Tale

Shelby needed a vacation … badly. She worked in the banking industry at a time when the economy was struggling and cutbacks in workforce were common. She wanted to get away but it seemed that none of her friends or family could afford the time or the expense of going on a trip.

Rather than just stay home and watch television and eat bonbons, Shelby thought this would be the perfect opportunity to take a little solo road trip to California. It was the height of the fall season, and the changing leaves in the mountains would be gorgeous. The temperatures in Lake Tahoe were predicted to be in the 60s, very comfortable for hiking, fishing, boating, and relaxing by the lake. Shelby made the decision that all signs pointed to this being the perfect opportunity for her to try out traveling by herself. She made her hotel reservations, prepped the car for the trip, packed her bag, and she was off. Her adventure was about to begin.

The drive was just as beautiful as she had hoped. The Rocky Mountain aspens didn't disappoint, and the sun was shining in the bright blue sky. It was perfect weather for a long drive. Shelby cranked up the tunes on the radio and sang her way across the Rockies. Life was good.

Shelby had packed a few snacks for the road, but when evening came she felt like she needed a more substantial and healthy meal, so she started looking for someplace to stop for dinner. She was about halfway by that point and thought it would be a good place to stop for the night as well. So, she found a hotel that the travel club had given good reviews and got a room. She inquired at the desk about a nice place for dinner, and the desk clerk heartily recommended the steakhouse down the road. Shelby was set. She carried her bags to her room, freshened up, and headed to what the clerk had assured her would be a great dinner.

The steakhouse looked promising. It had the quaint old wood decor of a good old mountain eatery with plenty of cars in the parking lot, always a good sign. Shelby walked in the door and was greeted by the hostess, who cheerfully asked, "How many tonight?" Suddenly, for the first time all day, Shelby felt lonely. She had never eaten at a restaurant alone before. Sure, she had done the fast food thing but had never sat down by herself in a crowd of people for a full meal. This is weird, she thought. "Just one," she told the hostess, and was then led to a table on the outer edge of the room. Shelby was grateful for that. She could sit on the perimeter and watch what was going on rather than being in the middle of the room feeling like a spectacle. She thanked the hostess and took a chair.

In seconds, the waiter was at her table with a basket of bread, a glass of water, and a welcoming smile. Shelby relaxed a bit and ordered a glass of wine. That would help her feel more comfortable. As she surveyed the room, she realized that she was the only person in the restaurant that was alone. Not one other table had less than two people. That glass of wine was sounding better all the time. Shelby was committed to making this work and focused her attention on ordering, eating, and doing everything in her power not to look up from her plate. That would only make me feel worse, she thought. She finished her dinner in record time, swallowed the last of her wine without really tasting it, and headed out the door.

On the walk back to the hotel, Shelby started thinking about her restaurant experience. Why had she felt so awkward about saying that she was alone when asked about how many at the table? Weren't people on their own entitled to eat in a restaurant too? The waiter had seemed friendly enough and no one had made any comments indicating that there was something wrong with being there alone. She suddenly realized that, in truth, she was the only one who felt awkward or thought it was strange. Nobody else probably even noticed or cared that she was alone. In that moment, she vowed to herself that for the rest of the trip she was going to try not to let being on her own cause her to feel bad. She would embrace this brave new Shelby and enjoy this independence she had decided to try on. Yes, this was *her* trip. She would hold her head high, bask in her bravery, and charge forward with confidence.

She proudly marched up to her hotel room door, turned the key, and ... "Uh, I'm alone again in this hotel room," she thought. She locked the door, sat on the end of the bed, took a few deep breaths, and said, "I can do this!"

That road trip was a turning point for Shelby. She experienced moments of true bliss, other moments of utter panic, but mostly moments of quiet reflection and self-exploration.

By the end of her road trip to Lake Tahoe, Shelby knew without a doubt how she felt about traveling alone. She was resolute in her decision, and has followed her heart and held to that decision ever since.

Pack Your Bag

All of us make many, many decisions every day. Some are pretty easy: what kind of breakfast cereal do I want, should I wear the blue shirt or the gray one? Others are a lot tougher: which house should I buy, do I want to change jobs? These I describe as low-cost and high-cost decisions. Breakfast cereal and shirts, in this case, are low-cost. If I choose one and it doesn't work out it's really no big deal. On the other hand, buying a house or changing jobs could have significant impacts on my life. These are the decisions that could cost dearly financially or emotionally, therefore I call them high-cost decisions.

Before we try anything new, we also have to take a quick look at ourselves, at the choices we have, and weigh the pros and cons of our choices. It is difficult to make any decision if we don't know why we are making this choice, what choice we can live with, and what choice would be absolutely unacceptable. This applies to ordering in a restaurant, choosing a mate, or buying a home. It is no different with deciding to travel alone.

Traveling solo is a choice that impacts you in many ways. Think of all of the decisions you have already made today. Your decisions are often made on the spur of the moment. This decision about traveling alone must not be spur of the moment, so spend a few minutes thinking of all the great decisions you have made in your life. From the time you were a small child, through your school years and into adulthood, you have made lots of good decisions.

List several of the very positive decisions you've made in life. Write down the choice and then briefly describe what the payoff or benefit of making that good decision was. You might have a little pain in your shoulder after this exercise from patting yourself on the back, but that's OK. You'll get used to it as you travel with me on this journey. I'll have you congratulating yourself often. All through this book I will provide examples when I ask you to write something. I have put them in a font that looks like handwriting so that you get a better idea of how to proceed. Remember, no right, wrong, good, bad, passing, or failing here.

Learning the Way

When you decide you might be interested in traveling alone, it is like deciding to try an exotic new dish at a restaurant. You have high hopes, but are not sure if you will like it, if it will be worth the price you will pay, or if you might just be all right with having the same dish that you've always had before. The truth is, either one is fine. It's up to you. But, what if that dish turns out to be the most amazing thing you've ever tasted in your life and introduces you to a whole new world of joy? It's worth a try, right?

Growing up, my mother was one of those moms who could make anything seem fun. She has always had a positive outlook and has never let her circumstances or anything at all stand in the way of a good time. One of my very fond childhood memories is of the two of us driving to the laundromat with a car

A Great Decision I Made...	The Long Term Reward I Got From It
Applying for the coaching position.	*I learned a lot.* *I met some well-respected leaders.* *It led to my administrative job.* *I was much happier with work life.*

full of dirty clothes. In my snotty adolescent voice, I asked, "Why do I have to go with you to wash clothes? It's sooooo boring." "No, it's not," said my mom. "It's an adventure!" Once I finished rolling my eyes and sighing heavily, I asked her what she meant. She then explained. She told me how life gives you what you look for. If you are going to the laundromat and expect to be hot and bored and tired, then that is exactly what you will be. If, on the other hand, you look at it as an adventure and look for the surprises, the interesting details, and try to enjoy your surroundings, the whole experience will feel like an adventure. That explanation shaped the way I look at not only traveling but at life.

It's like learning to ride a bicycle. You think you want a bike so badly. You finally get the bike and it's time for a grown-up to take you out for your first lesson. I don't know about you, but I was scared to death. My dad was one of those anti-training wheel guys, so straddling my two-wheeler, I started out. I wobbled and tipped. I wanted to give up and go home, but then I realized that each time I got back on, I got better. It became easier to stay up on that bike and eventually I was able to go all the way around the block on my own without even falling once. I was so excited! The world was now my oyster and I could go anywhere on my blue Schwinn. Tell me then, why do we know that we need to take baby steps to learn to walk, fall a zillion times to learn to ride a bike, but somehow believe that when we set out alone on a trip we are either good at it the very first time or we have to never do it again? I can hear you now. "No, no, that's silly. Of course, we need practice." Nobody gets it all right the first time.

I didn't start learning to ride a bike by signing up for a 100-mile marathon. I rode from one driveway to the next. I made the decision to quiet the fearful thoughts in my head and push through it. I made the decision to get back on after I fell off. Each attempt, each good decision led me closer to success. All I needed was to be given the freedom to flail. Notice I said *flail*, not fail. You need the chance to mess up and then the chance to do it differently the next time.

As we continue through this journey together, I completely understand that doing some of the exercises may be uncomfortable, difficult, or even downright scary. Think of all of the decisions you have made that were difficult. Many of those things turned out well, and I would bet that those decisions at some point were very stressful. It's human nature to have fear of the unknown and to be uneasy about making the wrong decision.

Deciding to travel alone is a big decision. It will definitely come easier for some than for others, but please remind yourself gently and often that you are new to this, that it takes practice, and that you will get better each time you venture out. It's just like Mom said: "If you look at something as an adventure and look for the surprises, the interesting details, and try to enjoy your surroundings, the whole experience will magically become much more fun and interesting."

So, be patient. Give this process the time and attention that learning a new skill requires. Try. Practice. Learn. *Then* decide. The best decisions are always informed decisions.

Adventure

It is often both helpful and interesting to have a conversation with someone you feel comfortable with and have them help you think about your reasons and desires for traveling alone. This person can become an ally and cheerleader for you as you work on getting ready to solo travel. Discuss the

2: Making the Decision to Go Solo

questions and topics below. Take the time to have this discussion to help you truly understand your hopes for your journeys alone.

Choose a friend or family member who knows you fairly well and who can be gently honest with you as you have an important discussion. Sit down with this trusted person and have a discussion about your thoughts on the travel-related questions below.

This adventure is intended to help you focus on all of the positive reasons you are thinking about traveling on your own. It is a way of helping you be clear about why you are considering solo travel, what you want and need from the experience, and what is definitely a deal breaker for you when traveling alone.

Your reasons and thoughts here should be your own and must not be influenced by anyone else's thoughts or needs. There is a point in time when you will need to consider the needs of others, but for now, this is all about you. When you are traveling alone you must depend on yourself, trust yourself and be honest with yourself. Otherwise, the experience becomes a negative one because you are taking someone else's journey, not your own.

I have provided space for you to jot down any thoughts or notes that you may want to save. You may choose to record your thoughts or you may choose to focus on the conversation and write none at all. It's up to you. So grab a favorite beverage for you and your friend and let the discussion begin. Enjoy!

Questions to think about for the discussion:

- Why are you thinking about traveling solo?
- Is it something you've always wanted to do?
- Is this a new idea for you?
- How did this idea come up for you?
- How are you feeling about traveling solo?
- What excites you about traveling solo?
- What worries you about traveling solo?
- Who do you know who travels alone, and what are your thoughts about their travels?
- What have you heard about traveling solo?
- Do those around you know that you are thinking about traveling solo? What have they said? Are they supportive or critical?

- Who in your life knows you are thinking about traveling solo and would be supportive of your efforts?
- What is your previous travel experience?
- How do you feel about experiences that are new, different, and unfamiliar?
- How do you react to change, being unsure, or being spontaneous?
- Describe your level of comfort with taking risks.

Things to consider:
- It is sometimes a little more expensive to travel solo. Do you have the financial resources to travel solo in a manner you would be comfortable with?
- How comfortable are you with being alone for long periods of time?
- Are you more of a social person or someone who enjoys doing things alone?
- Are you more of an extrovert or an introvert?

Notes/Thoughts:

Looking in the Rearview Mirror

Where is your Dream Destination? Get specific. Don't just say, "The ocean" or "Europe." Choose a particular ocean or part of the world. Europe is big. There is a great deal to see and do. Pick one particular country or city for now. You are going to do a lot of thinking, planning, and writing about this destination throughout the book so I want it to be somewhere you are truly excited about. That way you can enjoy the journey of exploring your wants, needs, interests, and fears. Dream big. The bigger the dream, the bigger the reward.

My Dream Destination is:_____

I have chosen this destination because:

_____
_____
_____
_____
_____

Now answer the following questions as honestly as you can. This work is for you. The clearer you can be with yourself the more likely you are to be making good decisions about your travels. Remember, no one will see this except you unless you choose to share. There's no grade for this, you have much to gain by being completely open with yourself, and at this point absolutely nothing to lose. Ready? Let's do this!

What is your reason for wanting to travel alone?_____

What are you hoping to gain from the experience?_____

_____
_____
_____

♦ Journey for One ♦

_____
_____
_____
_____
_____

What do you think has been keeping you from traveling alone before now?

_____
_____
_____
_____
_____

Do you know people who have traveled alone successfully? In your opinion, what makes them successful at it?_____

_____
_____
_____
_____
_____
_____

2: Making the Decision to Go Solo

If you had a magic lamp and three wishes for yourself related to solo travel, what would those three wishes be? _____

What skills, beliefs, and attitudes do you already have that you think would serve you well when you travel on your own? _____

♦ Journey for One ♦

What are your biggest concerns about traveling alone?_____

_____

_____

_____

_____

What are your beliefs about overcoming or minimizing those concerns? Can you do it?_____

_____

_____

_____

_____

Celebration

Congratulations! I'm proud of you. Looking in the mirror isn't always easy. There's nobody looking back at you saying only the things you want to hear. But boy is it worth the effort! You just took a big risk. You looked yourself in the eye and said, "Here I am. I'm not perfect but I certainly know myself well enough to know what I need and what I want. Bring it on!"

Your celebration today is something I think you are going to enjoy. You get to practice congratulating yourself for the risks you take, the challenges you've met, and the progress you are making. It's time to make a choice just for you. For this celebration, you are going to take a little mini journey. I want you to think of three places you have wanted to visit in your city, area, or a location nearby. Choose places that are new to you or that you are not very familiar with. Your celebration is to visit the three places, on your own, and enjoy yourself. That's it. Easy, right?! What a great opportunity to try out your decision-making skills and get to make your first true solo Journey for One. Now go out and make it a fantastic journey!

3

KNOWING YOURSELF

It's your road, and yours alone. Others may walk it with you but no one can walk it for you.

—Rumi

Travel Tale

Ruth was a woman who had always dreamed of traveling on a vacation other than to see relatives. Her dilemma was that her husband not only didn't like to travel but refused to travel. Ruth had lots of friends who had traveled a great deal and often told her of their adventures, amazement, and fun. Ruth desperately wanted to see Hawaii. It was a dream of hers ever since she could remember and she would have done *anything* to get to go. Anything, that is, except … go alone.

Ruth hated being alone. She and her best friends had weekly nail appointments on the same day and time so none of them would have to go alone. She would starve before she would consider going into a restaurant by herself. But she really, really wanted to go to Hawaii. So, Ruth's decision was that she would ask her sister-in-law, Jacqueline, to join her for the trip. The interesting thing about this is that Ruth didn't particularly like Jacqueline but she reasoned that going with someone she didn't like too much was better than not going at all. Tickets were bought and the countdown began.

Ruth was feeling that crazy combination of sensations that include excited, terrified, nervous, relieved, hopeful, and giddy the night before she left. She had made up her mind that she would make the most of her time in Hawaii and enjoy every moment. Her family got her to the airport and wished her farewell, happy for her that she was getting to have her dream.

When Ruth returned from Hawaii she was uncharacteristically guarded about sharing the details of her adventure. She would say that she had enjoyed her trip, liked the food, and thought it was a beautiful place. But the look on her face and the tone in her voice did not lead anyone who knew her well to believe that this had been the trip of a lifetime.

A couple of weeks after she returned from her trip she finally broke down and shared with a couple of family members that the trip had been terrible. She didn't get to do anything that she wanted to do, she had spent a lot of time in her room, and she almost wished she hadn't gone. After some questioning, her family found out that Jacqueline had been less than the ideal travel partner. The only thing she wanted to do was shop in the tourist areas. She had not wanted to do any sightseeing or participate in any activities. Ruth, scared of doing anything alone and not wanting to make waves with her sister-in-law, did whatever Jacqueline wanted and gave up all of her hopes and dreams for her trip.

For years after that, whenever anybody would ask if she had ever gone to Hawaii, Ruth would answer, "No." She was incredibly disappointed with the trip and with herself.

In later years Ruth's daughter-in-law gave Ruth the gift of a lifetime and presented her with another trip to Hawaii. This time Ruth got to do all of the things that she had dreamed of and got to do them with someone she loved. It was a tremendous gift to Ruth but also a good lesson for her family. A lesson that would later change the life of her granddaughter. Ruth was my grandmother. Without her experience, I would never have had such a passion for solo travel or the motivation to write this book. She gave me so much insight, courage, and determination. I am so grateful.

Pack Your Bag

Life is full of moments that are even better or even worse than we had hoped for. Traveling is no different. To have a successful trip it's important to know exactly what you like and don't like, your hopes and fears, and your must do and won't do experiences.

Before we dive into the lesson and this adventure, I'd like you to take a few moments to think about your preferences and fill in the chart below. Be as specific as you can without getting nitpicky! You don't need to explore that you like peas or carrots, but it is important to know whether or not you like heat, or jungles, or boats. Do you prefer an active vacation or something more relaxed and reflective? Do you have a preference for the cities or the countryside? Do you enjoy exotic or familiar? The more you can identify your likes, the better your chances for a great adventure.

Things I like to SEE on a trip.	Things I don't want to SEE on a trip.

Things I like to DO on a trip.	Things I don't want to DO on a trip.

Things that are REALLY important for me on a trip.	Things that I REALLY don't want when I'm on a trip.

My greatest HOPES for traveling solo.	My greatest FEARS about traveling solo.

Feel free to add to this list as you proceed on this and other chapters in the book. Again, the more information you have about yourself and your preferences, the more likely it will be that you will plan the trip of a lifetime!

Learning the Way

I want you to close your eyes for a moment. NO, WAIT! I mean *after* you read this. I want you to imagine that you have won a contest. The prize is a trip to the place you have always dreamed of traveling. The catch is that it's a ticket for only one person. You know how much you want to go and you decide you are willing to go alone if you can do everything that you've always dreamt of doing there. What are the things you've dreamt of doing?

When I started traveling I thought traveling meant I would go to a place, get on a tour bus and see whatever it was that the tour company deemed important for me to see. After a few trips, I realized that I had interests that weren't always what the tour operator deemed valuable. About three cities into a group tour of Europe I realized something important about myself. I liked giant, old cathedrals just fine. But, after about the fourth one with a long-winded guide, I had had enough. I felt like if you've seen one ancient cathedral, you've seen them all. I just couldn't see another one. Some of my traveling companions and I began to refer to each church stop as A-B-C (another bloody cathedral). It was one of my very first insights about my preferences when I travel. Another insight came after spending several hours with the group at a renowned art museum. I had seen the art in the way I wanted to see it in about 30 minutes. After that, I was in what I call the travel black hole. I had seen too much and I just couldn't see anymore. Just as with the cathedrals, I valued the work, respected the artists, appreciated the tremendous effort, and was in awe of its beauty, but I was able to enjoy it in my way much more quickly than many other visitors.

On the other hand, I also realized on this trip that the tour never covered some things of tremendous interest to me. I wanted to see a typical neighborhood and how the residents lived, where they went to school and where they did their shopping. Since I was a teacher, I wanted to visit a school and see if there were new ideas I could take away or stories I could tell the students about my country. This option did not exist on a tour.

When we travel on our own we get to make the decisions about what to see and what not to see. That's not to say that we can't or shouldn't do those other things, but if we have an interest we should do what we can to see or do it.

I'm not a hiker or climber and I'm not interested in extreme sports, but I do love to cook and do arts and crafts type activities. The answer for me is to research before I go somewhere the types of folk art or cooking that are typical there and find a course or teacher that might help me learn to make an item or cook a typical meal. I get to experience what I want. It just takes a little time and effort to make it happen.

I have come to understand myself much better through my travels. I know my comfort level with risk. I know I don't like crowds, fancy restaurants, museums, or visits to places that make you take a tour first and then end up at their gift shop. I absolutely love a trip to a grocery market, sitting to people-watch in a park or central plaza, visiting an artisan workshop, listening to local music, and anything to do with water. When I plan a trip for myself I get to make sure that I do the things I love, not what somebody else has decided I should love. This knowledge of myself and my needs has led to a much more fulfilling travel life.

With all of that said, I also make it a point when I travel to find something that takes me out of my comfort zone a bit. I remember a trip several years ago to Australia. I had signed up for a boat tour out to the Great Barrier Reef. I knew that scuba diving or snorkeling when we got there was an option, but I also knew that being seen in public in a swimming suit was not an option I was open to. Once we got out to the reef and I saw how happy people were and the amazement they were showing as their heads popped up out of the water, I decided to put my vanity aside and take the risk. I wanted so badly to have the underwater experience and knew that I would always regret not doing it. It's an experience that changed my life. I gave up my vanity, took up snorkeling as a hobby, and was unbelievably lucky enough to swim in the reef before climate change and tourism changed it forever. It just took a little bit of tough love, self-talk, and risk taking to make it happen. I will forever be grateful that I did. It's up to you. It's for you to decide what you want from your travels and how much you are willing to explore, change, and risk.

I often think, after I have talked to people who are fearful of traveling solo, that part of the fear comes from not knowing themselves and their needs, strengths and desires well. They aren't sure what to do and how to find out what else there is to do in a place. We are so lucky to have the internet and so many blogs and websites to help us plan and discuss our plans. I truly believe that this part of the process and knowing yourself well enough to know what you want and what makes you happy are the keys to successful solo travel. It's similar to being invited to an amusement park, given a full-day pass and then standing there trying to decide what to ride, what to try, and what you absolutely will not ride or do. It's up to you to give time, thought and understanding to what your desires are and what would make the visit the best for you. Whenever possible, this should be done well before you leave home. At the end of every one of my travels I like to think like Frank Sinatra. I want to be able to say, "I did it my way."

Adventure

This chapter's adventure is one that allows you to dream and dream BIG! I want you to think about Ruth in the opening story. You have the chance to go to your Dream Destination. Think about where that would be. Think about what you would like to see, do, and experience on that trip. Make a brainstorming list in the space below. Use every resource you have at your disposal. Talk to friends, look online, go to the library, watch videos, flip through travel magazines or brochures. Gather all of the information you can about what there is to do at your Dream Destination. Don't forget to think of things that wouldn't necessarily appear in a travel ad. I have a friend who enjoys going to a hairdresser and getting a haircut at her stops. Do you want to do something that's off the beaten path? Quiet, reflective time is also a valuable experience, so don't undervalue its place on your list.

♦ 3: Knowing Yourself ♦

Brainstorm List: Write down *every* idea you have of what you would like to do, see, or experience no matter how wild, crazy or seemingly impossible it might seem. This space is for recording each and every thought you have. You can't have your dream unless you have completely explored the possibilities for that dream.

_____   _____
_____   _____
_____   _____
_____   _____
_____   _____
_____   _____
_____   _____
_____   _____
_____   _____
_____   _____
_____   _____
_____   _____

Once you have your information as complete as you'd like, take the time to fill in the chart below about your destination. Try to prioritize and choose your top three to four choices for each area. Don't forget to look back at your Pack Your Bag activity at the beginning of this chapter to see if you are including (or not including) the things you wrote there. At the end of this adventure you should be much clearer about what your dream vacation would look like, as well as what it would take in terms of planning for you to have your most satisfying experience on your prize-winning vacation. I have included an example for a trip to Colorado next to each topic just to give you an idea of how it could look.

• Journey for One •

Dream Destination: *Colorado*
Things I'd like to see: *The Rocky Mountains.*
Things I'd like to do: *Go river rafting.*
Things I'd like to experience: *Riding a horse.*
Where I'd like to stay: *Log cabin preferably next to a river or lake.*
Form(s) of transportation I'd like to use: *Fly on a plane to get there. Rent a car to travel in the mountains. Ride the steam train in Georgetown.*

• 3: Knowing Yourself •

Things that are important to me: *Finding out about the Colorado Gold Rush.*
My greatest hopes for this trip: *I get to see some wildflowers and I become more confident in traveling alone.*
My greatest fear about this trip: *I'll get lost in the mountains.*
Thoughts/ideas I want to explore: *Could I go on a wilderness vacation alone?*

Looking in the Rearview Mirror

After you have completed the chart above as fully as you can, take a few minutes to reflect on the thoughts, insights, and attitudes you came to understand about yourself from this exercise. What are some things that you might want to take note of that would help you down the road in planning a real vacation? Please jot them down now. These notes are meant to be your memory bank so that one day when you decide to travel on your own you will remember the thoughts and feelings you had today and use them to help you plan your trip in the future.

_____

_____

_____

_____

_____

_____

_____

_____

_____

_____

_____

_____

Celebration

Unlike Ruth, who at the beginning of this chapter did not take the time to assess her own desires and needs and take responsibility for her Hawaiian vacation, you have taken a very important step in planning for success. You clarified for yourself what a journey needs to look like for you to feel successful and enjoy this wonderful opportunity.

Congratulations! Now celebrate by finding some little photo, memento, snack, song, movie or book that reminds you of this Dream Destination. Place it somewhere that you will see it often and then relax for a while and be proud of this very important step in your Journey for One.

4

THE FIRST STEP

Never underestimate the power you have to take your life in a new direction.

—Germany Kent

Travel Tale

Some might call it a midlife crisis. I call it a midlife calling. I was 50 years old, in the throes of menopause, and had been working in the same school district for 30 years. By almost anyone's standards I had a great life. I was single, owned my own home, had a stable and good-paying job, friends, and a family that loved and supported me. But I was restless. I was bored and just wasn't enjoying going to work anymore. I was fortunate enough to find myself eligible to retire and consider doing something else.

I knew that I wasn't ready to sit home and knit, and I kept feeling this deep desire for something more. I longed for something different, something special, an adventure. I had traveled a lot and loved being in new places, seeing new cultures, and having new challenges to conquer. I had to figure out what I wanted to do with the rest of my life.

I called a friend and announced that I had turned in my retirement papers. After he cheered and congratulated me for a bit he asked what I was going to do. At that point, I had no idea what I was going to do, but loved being in a position to have the whole world of options open to me. Before the end of the call he suggested an idea that would change my life. It was an idea I had never considered but immediately grabbed ahold of with both hands and my whole heart. My friend suggested I teach overseas.

I had always wanted to learn German because of my wonderful German grandfather. Maybe I should consider Germany or Austria, I thought. With the help of my friend, a pretty diverse resume, and a great deal of luck, within two weeks I was on my way to Vienna, Austria to interview for two jobs. Eighteen hours after arriving in Vienna I had a job and an apartment. My job teaching at the American school in Vienna would start six months later. What? Yes, it was that fast and that easy. I am not one to make hasty decisions, but once I make a decision I don't look back. I face the future with open eyes, open arms, and I expect the best.

As I began telling family and friends of my plans for the future, the responses seemed to fall into two categories. Some thought I was completely crazy, which most already knew to be true, and others thought I was incredibly brave. One day, words I'd never even thought about came out of my mouth in response to someone talking about my bravery. I replied that leaving the life I knew and moving to a foreign country was not brave. Staying put with nothing new, interesting, or wondrous in my life was brave. Brave because I would be risking spending the rest of my life being bored and uninspired. That scared the heck out of me. I have no idea where those words came from but they have become my motto and the guiding light for my life.

I arrived in Vienna on a very warm August day and can honestly say that the first couple of days were a bit of a blur. Between jet lag and the hustle and bustle of setting up a new apartment I had very little time to be anything but exhausted. Once I had my apartment looking something like a home, I began to settle into the reality of what I was about to do. I didn't speak the language, I didn't know my way around, I wasn't familiar with the public transportation, and I certainly had no idea how to navigate the endless list of chores I needed to do to establish myself in a new city. I spent hours and

4: The First Step

hours over the next several days working to figure it all out and doing my best to not feel like a complete idiot.

At one point, after a frustrating morning of bureaucratic mishaps and limited success in navigating the subway, I found myself walking down a main street in my neighborhood. Without warning I burst into tears. I was tired, frustrated, and feeling very, very incompetent. Again, out of nowhere, I thought to myself, OK Jodie. What is it that you need right now? What would make this situation feel better? I realized that I needed to feel normal, my normal. I needed to sit and drink a cup of coffee, read an English newspaper, and breathe. I needed to remind myself that beginnings are hard. It's part of the process. It's the first step that's the hardest. Once you get past that, things get easier. So, I stopped at the newsstand and got a U.S. newspaper, then sat at an outdoor cafe with a cup of coffee, and spent the next 30 or so minutes reading and breathing. I needed to let my brain rest. I had been working very, very hard for the last several days. I needed to respect the fact that even I had limits. I could only do what I could do. It worked. While I sat at the cafe becoming normal, I renewed my resolve, gained another dose of courage, and set off anew to face the challenge.

As I look back at that period of time, I am incredibly proud of myself. I gave myself time, encouragement, and the freedom to flail. Not fail, but flail. I gifted myself with the time and permission to struggle, make mistakes, and try again. It would later take a great many small victories, missed steps, and false starts but that first step into living overseas was the foundation for what turned into an incredible 11-year career and life in Vienna. It is without a doubt the best 11 years of my life. I found renewed energy for my profession, I met incredible people I now consider family, and I found myself. I had the opportunity to grow, learn, and find joy in things that I would never have known had I not taken that first, very big, very scary step into my future.

Webster's Dictionary defines bravery as "the quality or state of having or showing mental or moral strength to face danger, fear, or difficulty." Taking a first step toward anything new takes bravery. It means you have to face things you would probably rather not face, but believe me when I tell you that facing the fear, being brave, and taking the first step toward traveling solo is the biggest and most rewarding gift you will *ever* give yourself. Go ahead. Take the step. I'll be with you the entire way. A world of wonder awaits you.

• Journey for One •

Pack Your Bag

We all take many first steps in life. We take first steps when we learn to walk. We have first days of school, first dates, first jobs, first home, first children, and the list goes on. Think of all the firsts you have experienced in your life. You have certainly felt success with many of them and learned from both the successes and those that didn't turn out quite so well.

What are some of the important firsts that you have experienced that turned out well for you? What firsts encouraged you to go on and do even more than you dreamed you could? Write down a few and reflect a bit on how great it feels to try something new and succeed.

My first...	Reflections on the success of that first step. How did it feel? How did it encourage you to go on?
Example: diving off of the high diving board	I was scared to death but had encouragement from others so I did it. I was so proud of myself. I knew if I could do this I could do anything. Made me want to go on to learn to be a diver and I did.

Learning the Way

One of my favorite stories when I was growing up was *The Little Engine That Could*. In case you are unfamiliar, it is the story of a little train engine that was loaded with toys and was having trouble getting up a big hill. The little engine finally started saying to itself over and over again, "I think I can. I think I can." The little engine made it up that hill and learned a very valuable lesson. He learned that you can do anything you put your mind to. All you need to do is try.

Making the decision to buy this book and take the Journey for One is an important one. It means that you quietly, or perhaps loudly, said to yourself, "I think I can. I think I can." Just like the little engine, you have opened yourself up to the possibility that you have the desire and can develop the courage and skills it takes to travel on your own. You're absolutely right. You *can* become that courageous, and the skills are there for you to learn. All it takes on your part is a commitment to doing the work. That commitment to doing the work is just the first of many first steps you will take on this journey. Some of those steps will be small, others will be big steps. The important part is that you take them.

Taking first steps was explained to me in a meaningful way several years ago by William Bridges, a man who studies change. Bridges wrote a book called *Managing Transitions*. In his book, he explains that making a change, like traveling alone, requires a transition for us. It means going from one thing or one way that we know and transitioning or moving into something that we don't know ... yet. He says it's like stepping across a stream. Standing on one side of the stream with both feet firmly planted on solid ground is comfortable. It's safe. We feel strong. Once we are on the other side of the stream with both feet again firmly planted we again become safe and strong. It's that moment in time when we have taken that first step off of the bank and are suspended, momentarily, over the water that is the scariest. Will we fall in? Will we lose our footing when we step onto the other bank? What if we can't do it? What if we fall short? All of those questions are perfectly valid and arise every time we try to do something new or make a change. The secret to success in making a change comes with that "I think I can" moment. At some point, we have to have the courage to take that first step. We have to stretch and be determined that we can do it, that we *will* do it. Like Henry Ford, the automobile pioneer, said, "Whether you believe you can do a thing or not, you are right." It's up to you.

Luckily for you, taking this first big step will not require you to leave the shore, get wet, or be alone in your crossing. I will carefully lead you through the steps and show you how you can become someone who can and will be able to travel solo.

If you are still with me and are still willing to start down this road, then let's go and see what fun awaits us on your next solo adventure.

Adventure

This adventure on your Journey for One is a fun one. It's something that gives you what will become your very most important and prized possession. You are going to apply for and get your passport if you don't already have one. Your passport is the key that opens many doors when traveling. Our world has changed, and other than traveling within your own country of citizenship, there is nowhere you can go without a passport. What an incredible shame it would be to go through this book, complete all of the adventures and exercises, decide you are ready to take the plunge and schedule that journey for yourself — only to find that you can't go because you can't get your passport in time.

For those of you who already have a passport, make sure that it has more than six months left on it before it expires. Most airlines will sell you a ticket with or without a passport, but once you arrive at the airport, the airlines will not let you board the plane with less than six months left until your passport expires. Trust me, I know this situation well. We had three people who almost missed my niece's wedding in Italy because of this situation. It is a costly and frustrating issue that is best avoided.

If you are a U.S. citizen in good standing, getting a passport is a relatively smooth process. Below, I have explained the various ways to proceed depending on your citizenship and situation.

If you are a citizen of the United States of America and this is your first passport, go to www.usa.gov/passport online and follow the process for obtaining a first passport.

If you are a citizen of a country other than the United States of America, go online and search for the government agency that takes care of passport matters for your country of citizenship and follow the directions they give for your situation.

If you need to renew your U.S. passport, go to www.usa.gov/passport and follow the instructions for renewing or replacing a previous passport.

All right! The journey has begun. This step is an important rite of passage for travelers. Your passport opens doors, increases your options, and is a tangible symbol that says, "Hey, look at me! I'm a traveler."

Congratulations traveler. Welcome aboard.

Looking in the Rearview Mirror

Reflecting on steps we have taken to reach a goal is important to our learning. It's important to make note of what we did, how we felt, and how things turned out. It may not mean much at the time, but as we continue the journey our memories tend to fade and we forget where we started and how much progress we have made. A time will come when getting a passport will seem like no big deal, but for now it can be very symbolic and meaningful.

Take a few minutes to reflect on your process of getting a passport. Consider how it felt to take this step. How did the process of applying for the process go for you? What thoughts did you have as you considered having a passport? What hopes do you have for your passport? What countries do you hope you get to have stamped into your passport? Jot down any other thoughts you had about getting or renewing your passport. If you already have a passport, write about some of the places your passport has taken you and what places you intend to visit that will someday be stamped in your passport as a solo traveler. Happy writing!

_____

_____

_____

_____

_____

◆ Journey for One ◆

_____

_____

_____

Celebration

For this celebration, you get to pretend that you have all of the money you could ever use and every wish you would ever have would be granted. Use whatever resources you have at your disposal to make a wish list. This will be a list of all of the places you think it would be fun, interesting, or even magical to visit. The list doesn't need to please anyone but you. If you think visiting the prairie dogs on the plains of the Midwest sounds like fun, then write it down. Perhaps it's an exotic jungle trek that makes your heart go pitter pat. You are the master of your wish list. No judgments, no fears, no excuses: just write them down, one and all. Ready, set, go!

My Travel Wish List:

_____

_____

_____

_____

_____

_____

_____

_____

5

SETTING YOUR GOALS

The purpose of life is to live it, to taste experience to the utmost, to reach out eagerly and without fear for newer and richer experience.

—Eleanor Roosevelt

Travel Tale

Several years ago, I had the opportunity to travel to India for an international teachers' conference. It had long been a dream of mine to see India and experience the sights, people, and culture so very different from my own. This time the travel fairies had sprinkled magic dust on my life. Not only was I going to India but the trip was being completely paid for by my employer. I couldn't wait. There was so much to do and see, and I knew I wouldn't be able to see it all. No matter, I would see as much as I possibly could.

During the *very* long flight to Delhi I took out a notepad and pencil and began to make a list of the things I really wanted to do or experience while I was there. It started out with the usual things like the Taj Mahal and a souk market, and then slowly began to include simpler or more ordinary things that I wanted to experience. For example, I knew that cows were sacred in India and that they were allowed to roam freely anywhere in the country. I

◆ Journey for One ◆

truly hoped to see a cow in the middle of the road stopping traffic. I wasn't a big fan of curry but thought I should do my best to experience different kinds of Indian food and develop at least an appreciation of it. I also felt a need to try on a sari. I had worn one that belonged to a family friend as a little girl, but wanted to try it as an adult and really learn how to wear them. So, as you can see, my list included the typical tourist things, but also little personal challenges. Unbeknownst to me, I had just made my first ever travel goal list. After I had landed, I looked at the list and knew that I had a plan. I didn't need to wander aimlessly around the country or be tied into a group tour that wasn't getting me what I wanted from the trip. I was free to make this *my* trip, see what was important to me, and come away from the journey a more culturally developed person. I dug out that old India list from my memory book the other day:

- See the Taj Mahal.
- Eat a different Indian food daily.
- Ride a tuk-tuk.
- Visit a souk market.
- Learn more about Gandhi.
- Try on a sari.
- Experience an interaction with a sacred cow.
- Learn to say the words hello, please, thank you, and goodbye in Hindi.
- Experience something that I had no idea to plan for.

After I had gotten to my hotel, cleaned up a bit, and felt like I was ready to look around a little, I went to the hotel restaurant for some lunch. I was eager to sort through my travel notes and browse the many tour brochures I had collected from the hotel lobby. When the waiter arrived, and asked if I'd like to start with something to drink, I remembered my goal to try new things, so I ordered their house chai tea. When he returned with my tea, the waiter asked if I knew what I wanted to order. I explained to him that I didn't know Indian food well and that I was excited to try new things, so could he please suggest something. After a little round of 20 questions the kind young man knew that I didn't eat red meat, loved vegetables, loved spicy food, and that I had no allergies. About 15 minutes later he returned with the most beautiful and colorful plate of food I think I have ever seen. It was soooo ... good.

I wrote down the name of the dish but over the intervening years have lost the little paper. What a shame. I would give anything to have that dish again now. Anyway, after that first delightful meal I told the waiter that because my conference was in the hotel it was likely I would be in there quite often. I asked if he would be willing to choose my meal for me each time I came in. He seemed pleased to be given the challenge and I was thrilled at the possibility to make tremendous progress toward achieving my Indian food goal.

I have to admit that seeing the Taj Mahal was, for me, more of a "should do" rather than a "wanna" do. I don't know why, but I just wasn't interested. However, I knew that this could possibly be a once-in-a-lifetime opportunity and it is, after all, one of the Seven Wonders of the World. So, I checked with the tour desk in the hotel and asked about options for going the three and a half hours from Delhi to Agra to see the Taj Mahal. After considering the options I chose to hire a private car and driver, which also included an all-day private guide in Agra. This seemed awfully extravagant for me given the way I usually travel, but I wanted to be extra safe and careful since I was on my own in a place that felt, to me, to be in a location where a little extra precaution wouldn't be a bad thing. Besides that, the economy in India was such that this entire day was only going to cost me $100 U.S. What a wonderful decision it turned out to be. The driver was kind and had grown up in the area so was able to tell me stories about the towns and people along the way. He shared the yummy snack that his wife had packed for him to carry in his car. I found out that the tour guide's wife was a teacher so he offered to take a little side trip in the afternoon to visit the school where his wife taught. I ate at a special little restaurant, petted a monkey, rode a tuk-tuk, learned about the amazing history of the Taj and fell in love with that amazing wonder of the world. Add three checks to my "did it" list. I even got to make two checks in the "something I didn't know to plan" column for petting the monkey and learning the history.

I am so glad I convinced myself to visit the Taj Mahal. It was an experience I treasure. There were other wonderful experiences in India too. I did learn those Indian words, realized that my body was not made to look good in a sari, and sat at a dead stop for 35 minutes during rush-hour traffic in Delhi to wait for a sacred brown cow on the road to move. Frustrating? Absolutely not. It was one of my favorite experiences of the trip.

Pack Your Bag

Think about something that you once really wanted to do and felt like you would likely be successful at. Perhaps it was playing a certain piece of music or climbing a special mountain peak or even learning a new language. My guess is that you didn't start by playing " Chopsticks " with the short-term goal of playing Rachmaninov. We all have to start somewhere. It's like they say, it's not about the destination, it's about the journey.

Think about something you did when you were a beginner at it. Think about how hard you worked to become better. I want you to think about the steps or mini goals that you set for yourself and accomplished before you became what you considered good at what you did. Write down the steps you remember reaching and what they helped you do to get to the next step. I've given you a little example to help get you started.

Example Chart

What I worked hard to accomplish ...	Learn a Mexican folk dance.
What level of proficiency did I start at?	Not a great dancer and two left feet.
What tasks/steps did I work toward to reach my goal?	Found lessons at local rec. center. Took dance class. Practiced daily.
What did success at one level do to motivate me to take the next step?	Once I got the basic steps I wanted to use them to do a "real" dance. I wanted to keep going.
After taking those several steps what were you able to do?	After I learned the dance I performed with the class for a local Cinco de Mayo festival. Very fun!!!

Your Chart

What I worked hard to accomplish...	
What level of proficiency did I start at?	
What tasks/steps did I work toward to reach my goal?	
What did success at one level do to help me feel better about the next step?	
After taking those several steps what were you able to do?	

Learning the Way

My trip to India was everything I had ever hoped for and more. My waiter friend at the hotel introduced me to foods that I loved so much that I have now learned to make them at home. I was able to spend a day in a public market celebrating Gandhi's birthday. I was amazed innumerable times by the number of people the ingenious Indians can squeeze onto a motorcycle, something I never even knew I would see. In my view, I got my dream trip because I had planned the dream. I had set out goals for myself and created a list of things to see, do, learn, and experience, which all but assured my success in making it a trip I would love. I believe that traveling without a few goals makes it almost impossible that you will ever experience a truly satisfying journey. It would be like heading out on the open road with no itinerary and no map and then being disappointed at the end because you didn't go where you wanted to.

Goals are the fuel that gets us to do or be what we want in life. They give us direction, purpose, and clarity. Athletes use them to train for the level of proficiency they want. Businesses set goals for the financial outcomes they want. Schools set goals so that students learn the most and the best that they can. Traveling is exactly the same. When you have a goal, you have something to look forward to, focus on, and feel good about completing.

If you look for information about goals you will find a *ton* of information about how good they are, how to make them, why people don't achieve them, and the best ways to establish them so you are more likely to achieve them. I have found, after years of teaching and leading workshops, that everyone is unique. What works for one won't necessarily bring success for another. You need to figure out for yourself what works for you. If you write your goals down somewhere, does it make you more likely to achieve them? Some people say that telling someone about their goal supports them by making them feel more responsible and obligated to fulfill that goal. Yet others succeed best when they keep their goals to themselves so that if they fall short of their goal they aren't embarrassed or ashamed of themselves. Whatever you decide to do, share or not share, it's up to you. This step of the process is set up to ease your fear, get you what you want/need, and create a positive outcome for your travels. Please don't let a silly thing like sharing a goal become a stumbling stone to your absolute success and joy. Make the process work for you.

Another thing about goals is that they change. I may set a goal that I want to learn to Flamenco dance when I go to Spain. Let's say I get to Spain and as I'm searching for Flamenco lessons I stumble across a class in learning to click the castanets like you see in the movies. Oh, my gosh! I had forgotten that I'd always wanted to do that. My time and my wallet both say I can't do both so I decide that I'd rather learn castanets this time than take a Flamenco dance lesson. I decide it's more likely that I could find a Flamenco teacher at home than I could a castanet teacher. Done. Castanet lessons it is. No shame, no disappointment, no explanation necessary. When you have the luxury of traveling alone you get to do exactly what you want when you want. You don't have to explain your choices to anyone. It's like the line in the movie *To Wong Foo, Thanks for Everything, Julie Newmar*, "Your approval is neither desired nor required." It's OK, go ahead and change those goals as you go. It's your trip, your dream and your time to enjoy.

For some, a goal may be to be open and free with no plan, which is a marvelous way to go for some travelers. You know yourself. Do what's best for you. I would suggest, though, that even if you don't have a concrete plan for your time or destination, you still have goals. They might be things like noticing wildlife, or learning to trust your gut, or going with the flow of the journey rather than the ticking of the clock. Do it your way. It's up to you to guarantee the success of your solo journey.

Adventure

What is your idea of a perfect solo journey to your Dream Destination? If you could custom design a trip and fear wasn't a factor, where would you go and what would you do at that location? Write down the name of your Dream Destination, then think through the following questions. Research your location, using the resources of your choice to find ideas for what you might like to do, see and experience there. Jot down some notes. I encourage you to look back at Chapter 3 so you don't have to start from scratch (unless you want to, of course).

- Where would you like to go? _____

_____

_____

_____

♦ Journey for One ♦

- What would you like to see? _____
  _____
  _____
  _____

- Where would you like to stay? _____
  _____
  _____
  _____

- What kind of transportation would you want to take? _____
  _____
  _____
  _____

- How and what would you like to eat? _____
  _____
  _____
  _____

- What would you like to do? _____
  _____
  _____
  _____

- What would you like to learn? What specific experiences would you like to have? _____
  _____

○ How long would you like to be away? _____

○ How connected would you want to be with your home life? _____

○ What kinds of traveling would you like to do? (Adventure, beaches, shopping, nightlife, volunteering, ecological, hardship, spiritual, cultural, athletic, artistic, family/friend visits, farm stay, sport experience, hiking, biking, etc.) _____

○ Would you like to be alone or in a group? (Solo travelers sometimes join a group.) _____

• Journey for One •

○ What are your greatest hopes for being solo? _____

_____

_____

_____

○ What previous experiences have you had that might be similar to what you are interested in or that you could build upon? _____

_____

_____

_____

Now that you have identified most of these elements for your dream trip I want you to think about what kinds of things are on the list that might frighten you to do alone. Earlier when you were writing, and I asked you to think of it from the perspective of fear not being a factor. Now I want you to identify the things that you're pretty certain would frighten you. Let's say, for example, that one of the means of transportation that you'd love to experience is to go sailing in the open ocean in the Caribbean. This time when you think about it you know that doing that without a travel buddy, just the boat's captain and crew, would scare you. Then I want you to write it down on the list below. No judgment, no expectations, just write it down. See if you can think of two or three of the things from your dream trip that might be scary for you.

What I want to do that frightens me

• 5: Setting Your Goals •

Now take each one of those items that you want to do and create a goal for yourself. I have written an example for you to follow. Notice how I have broken the fear down into smaller pieces so that I can approach the challenge one bite at a time, not the whole thing at once.

What I want to do that frightens me…	My goal for this scary experience
EXAMPLE: Ride a camel in the Sahara Desert	○ Find information about how and where to take a camel ride. ○ Talk to others who have ridden a camel. ○ Go to a place that has camels for hire and talk to them about trips and options. ○ Get close to the camels. ○ Schedule a trip so that I will treat myself to this adventure.

Looking in the Rearview Mirror

You have worked hard on this chapter. There is a lot of self-exploration that you were asked to do. Those little journeys are more difficult for some than putting on hiking shoes and walking five miles. For others, this was invigorating, relieved stress, or gave you just the kick in the pants you needed to start turning the dream into action. Spend some time reflecting on how making a plan helped you, relieved you or perhaps scared you. Think about what it was that made you feel the way you do. What could you do to feel better about the feelings you are having?

If the experience of this activity was helpful and made you smile thinking of all of the possibilities, I want you to create a postcard to yourself. It can be the same size as a full sheet of paper, the size of a normal postcard, or any size that you want. I want you to draw, glue, paint, or somehow visually represent the destination of this dream trip you planned on one side of the card. On the reverse side of the postcard, where you would normally put the address, I would like you to jot down the two or three goals you had for yourself. In the section of the card where you normally write a message, I want you to write a positive message to yourself about how it felt to dream and begin to plan a very doable journey for yourself. Give yourself some encouragement, credit, and thanks for taking this big step toward making this solo travel destination dream a reality.

Celebration

Take the postcard you just made and put it somewhere that you are likely to find it in a month or so. Maybe that's a special jewelry box, maybe it's the sock drawer, perhaps a pocket in your winter coat. The location doesn't really matter. The idea is to celebrate that you were able to create a mini plan for yourself that gave you another strategy and a bit more courage to help move you closer and closer to that dream solo journey. Congratulations! Now go have something fun to drink that you have never tried before. Give yourself some congratulations for moving forward, taking a chance, and dreaming big. Cheers!

6

THE JOY OF SUCCESS

The more you praise and celebrate your life, the more there is to celebrate.

—Oprah Winfrey

Travel Tale

When you were a kid, did you ever lie on your back with your head over the edge of the bed and wonder what the world would be like if we lived on the ceiling instead of the floor? Be honest! My own personal research, based upon a random sample of slightly inebriated friends, suggests that most of us have, indeed, had that thought at some point in our lives. That upside-down view of the world is exactly how I felt about my upcoming solo road trip in Ireland. It would be like seeing life from the opposite perspective. I would be driving on the opposite side of the road and I would be driving on the opposite side of the car. I couldn't wait. It was going to be so much fun.

I was off to a great start until I stepped off the curb at the airport to board the shuttle bus and almost got run over because I forgot that the traffic would be coming from my right instead of my left. Oops! I'd better pay more attention, I thought, with the caution of a preschool teacher in charge of a field trip for 3-year-olds.

• Journey for One •

A couple of minutes later I smiled as I sat safely in the little van that would take me to my rental car for the next seven days.

The first thing I did after putting my suitcase in the back of my rental car was to throw my purse in the back seat and slide gleefully into the ... wrong side of the car. I'm in the UK, I reminded myself impatiently. The steering wheel is on the other side of the car. I won't embarrass myself by telling you the number of times I did that over the next week, but I will say that I took every opportunity to celebrate each time I remembered to get in on the correct side of the car. I needed all the positive reinforcement I could get.

I pulled out of the airport parking lot, remembering to look right first for oncoming cars. Yaaaay me! I cautiously starting driving and figuring out just the right spot for centering the car in the lane. This wasn't going to be so hard after all. Good job, Jodes! (my family nickname). I could do this. Each day I continued to improve, pulling into traffic and staying in the correct lane as I made my way around the beautiful green island. I was gaining skill and feeling pretty confident by the time I reached midweek. It actually started to seem more logical and instinctive by the time I left. Yes, I got honked at, pointed to, reminded, and laughed at each time I made a mistake but it seemed that the lovely Irish were used to people messing up while driving there. With absolutely no conscious planning, each time I made the right move while driving, I began wetting my index finger and making a tally mark in the air to award myself another bonus point for being successful and improving. My score never mattered. I never kept track, and I'd start over every day as I started a new adventure, but it gave me a way to acknowledge that I was doing it right. I was getting better and I was having fun. I didn't realize it at the time but I believe that this simple act of awarding myself points built my confidence and my comfort in driving that way. The next time I visited a country where I needed to drive on the left side of the road, I had a visual memory of being successful and having fun doing something that could have otherwise continued to cause me stress.

Since that time, I've had to learn other new skills while traveling. Celebrating *with* myself and *for* myself each time I do well is a way I continue to be kind to myself and have fun while learning. If I were traveling with someone else I would be saying, "Hey, look at that! I did it right!" My companion and I would hoot and laugh and we'd carry on. When you travel solo you get to do that for yourself. I don't leave this little ritual out when I'm traveling alone. It's too important. Feeling good about my new adventures and my successes is part of my joy in traveling on my own.

Pack Your Bag

We all like to do well. We like to succeed at those things we are trying hard to learn. Ideally, when we do something well the good feeling of having made that accomplishment can be enough. But, it can be even more fun and reinforcing to acknowledge our successes.

What is something that you might consider rewarding to do for yourself when you are successful at a new skill or accomplish a new goal? It is important to remember that it does not have to be anything complicated or expensive. As I told you in the story, sometimes it's as simple as a tally mark in the air. Some examples might be time to yourself, a long bath, or a nice walk. It can also be something social like time with a friend, coffee with a colleague, or a date night out with someone special. Give it some thought, do a little observing or research, and jot down several ideas or fun reinforcements so you have them on hand when it's time to celebrate!

When you want to pat yourself on the back for a job well done, it's handy to have some ideas of what might feel like a special reward or treat to celebrate your success. In the space below, list several things that you would enjoy as a reward to celebrate your accomplishments on your journeys. It is much easier to think of a reward when you have done the brainstorming ahead of time. Be creative and have some fun. Remember, it doesn't have to be big or complicated, just enjoyable.

Special Rewards I Enjoy

Example: *Enjoy a cup of tea while listening to music.*	

Learning the Way

We, as humans, are goal-oriented beings. We set our sights on many things in the course of our lives. We work towards graduating, losing weight, saving money, finishing books, organizing and cleaning spaces, and a million other things we view as important in improving our lives. One of the things that is part of our culture is to celebrate when we have met that goal and invite others to celebrate with us. We have graduation parties, baby showers, and wedding receptions. We reward ourselves and each other when we have done something noteworthy. We buy ourselves a new outfit or go out for a nice dinner when we've finally completed something we set out to do. We do this in order to say, "I worked really hard and achieved something important." It's a ritual, and psychology tells us that rituals are important.

I am a big fan of rituals. As I have grown as a solo traveler I have learned how important it is to provide myself with a ritual to celebrate the success of my travel triumphs. It is psychologically important to acknowledge and reward ourselves in the same light that we were celebrated when we learned to ride a two-wheel bike on our own or when we got our driver's license. Those are big milestones in life and the celebration of reaching those milestones provides the encouragement and excitement we use to push us to even bigger goals and aspire to more fulfilling journeys.

There are a multitude of ways to celebrate and reward yourself for a job well done. You don't have to spend a lot of money, or any money at all to treat yourself to a valued celebration. Over my years of travel my ritual celebrations have varied and changed. At one point in time I would collect a few recipes from my destination and then invite friends over to sample the food and see the slideshow from my latest journey as my way to celebrate my success, safe return, and tremendous joy. Later I found that a simple pair of earrings from my destination that I could wear and recall wonderful memories was the way to go. I've included good bottles of wine, stones, photos, games, and massages as ways to pat myself on the back for my courage, adventurous spirit, and cultural awareness. Everyone values different aspects of their travel and celebrates those aspects in very different ways. How you celebrate is not nearly as important as the fact that you *do* celebrate. Psychology tells us that there is not only a psychological benefit to acknowledging our successes, but that there is a physical benefit in even the simplest celebration. Even raising your arms above your head and yelling *yes!* is enough. It provides us with reinforcement for the positive attitude and efforts that we will want to have when we are faced with future adventures and challenges. It builds our self-esteem, our sense of competence, and our confidence to acknowledge a job well done.

Let's take a look at a fairly typical solo travel day scenario. You get up in the morning and find that you slept well despite the bed being different from what you sleep on at home. *Great. I'm off to a good start*. You dress, have breakfast, and start your day. You've slept a little later than you hoped but there's still a chance you can catch that 10 a.m. tour. You hurry off and get your ticket for the tour just as they are about to close the window. *Wonderful, yet another success*. After the tour, you want to go to a restaurant that was recommended by a friend at home but you're not sure where it is. Out comes the map, and after a couple of moments of weighing the options you choose your route and take off for lunch. After a few turns and bends on the cobblestone streets you become turned around and not sure where you are or if you're even going the right way. You figure you can keep wandering and hoping but you decide to stop and attempt to ask a shopkeeper if you are close. You show him the map and he smiles, waves you to follow him and leads you around the corner to the entrance to the restaurant. *Bravo, victory number three*. You order your lunch and add a favorite beverage to celebrate your problem-solving and navigation skills. After lunch, you decide you want to sit somewhere quiet and catch up on your journaling. The little place down the street with the fountain looks appealing. You spend time writing and get all caught up and then briskly close the cover on your journal with a big sigh and a smile to celebrate your success for staying current with your writing. Yes, that little snap, sigh, and smile are all mini celebrations of your accomplishment.

As you can see, even our relatively simple days can provide many opportunities for celebration and success. Each time you consciously acknowledge that something good has happened you build your traveler esteem and confidence. That's exactly what good coaches do. They plan the practice so that the athlete has an opportunity to try out new skills, praise their improving skill and success, and then remind the athlete that they can do this, they've done it before. The difference is that when you travel solo it's important to be your own good coach and encourage your own growth. This may seem silly and unnecessary to some to celebrate these little bits of success, but I assure you that what you focus on is what you get. If you focus on all the things that don't go well, that is all you will see. You will completely forget that you had a zillion little moments that went well and that you handled with tremendous solo traveler skill.

Did you ever participate in the Girl or Boy Scouts as a child? Do you remember the sashes they wore proudly across their chests that displayed rows of colorful badges for having successfully completed a set of skills? Once you earned one badge and mom sewed it on, you couldn't wait to start on another one. Each of those badges represented your efforts, skills, and achievement in navigating life. Just as we collected those badges as a symbol of our growth and competency, I encourage you to think of some way to commemorate and celebrate your travels. I suggest you find something that you can collect on your travels that allows you to reflect back upon your trip and think how wonderful and successful it was and how proud you are of yourself for taking the risks, following your heart, and trusting your gut. I've known people who have collected transportation tickets, brochures, photos, postcards, bells, masks, jewelry, buttons, pencils, and any number of silly tourist trinkets. There will come a day when you will sit down and pull out a box or bag of your little tidbits from your travels. It will bring great joy to your heart because of the wonderful memories it represents and it allows you to take pride in how far you have come as a traveler. Just as the scouts treasured those badges, we travelers should keep some little reminder of our accomplishments for ourselves.

In the high-pressure world that many people now live, it's easy to forget the many little moments of light that we experience every day. We need to take the time to see, acknowledge, and celebrate those moments of light for they are what can keep us moving toward our greater goals.

"The more I accomplish, the more I know I'm capable of accomplishing."
Tawny Lara

Adventure

Your mission during this adventure is to experience success. I want you to think of something that is a new skill for you. It might be travel-related, such as reading flight offers on an airline website, or something unrelated to travel such as crocheting or cooking. Choose an activity related to that skill that you can spend some time practicing. As you begin to practice, be aware of the steps when you are successful.

Let's say you are crocheting. As you begin, you know that you have never made a chain stitch before so you will pay attention to how you are doing on that step of the task. Each time you make a successful chain stitch you give yourself a high five or a tally mark in the air to celebrate your progress.

Once you have completed the task to whatever degree you want, reward yourself with one of the reinforcements from the Pack Your Bag sections of this chapter.

Looking in the Rearview Mirror

Each of us needs a different form and degree of reinforcement for our achievements. Depending on our personalities, level of skill, and a number of other factors, we may need more or less acknowledgment of our success. There will be times in your traveling solo journey that you do something that seems easy or like it is not stressful. You decide to what degree you want or need to celebrate. On the other hand, when you have accomplished a task that has taken a good degree of effort or courage on your part, I encourage you to truly celebrate that progress.

Go back through the earlier chapters in this book and reflect upon all that you have already done to become more skilled and confident in traveling solo. I'm sure you will see that some have been fairly simple, others more challenging. Jot down, on the chart below, where those tasks fall on the easy to difficult spectrum. This is so that you have a visual representation of how you are progressing on your travel journey.

Tasks that were easy for me	Tasks that challenged me
Listing the places I want to visit.	Thinking of ideas for celebrations.

Celebration

Take a look back at the adventures you listed above. Remembering how much effort and courage it took for you to accomplish that task, please choose a reward or several rewards for yourself and celebrate the successes you experienced in doing those tasks. You may choose something simple for some tasks and some rewards that are more detailed for others. Remember, each time we reinforce ourselves for a job well done, the better the likelihood that we will be successful in undertaking that task again. Go ahead. Celebrate yourself and the progress you are making. You have made big steps on this Journey for One. Bravo!

Task I accomplished	How I celebrated
I renewed my passport.	I bought myself a bottle of wine from my dream country.

7

WHEN THINGS GO WRONG

Sometimes things go wrong and no matter how hard you try, you just can't make them right again. That's when you accept the reality of today and plan the strategy for tomorrow.

—Author Unknown

Travel Tale

Joan, a trauma therapist, was invited by a local university to be a part of a therapeutic debriefing team to go to Russia during the aftermath of the Azerbaijan conflict. She was an experienced international traveler and was thrilled to be going as part of a volunteer team to provide emotional care to the victims of the violence. She and her team knew that because they were coping with a war zone situation that their accommodations and creature comforts would be fairly minimal. They were mentally prepared for anything.

After arriving during the night, hauling their suitcases up three flights of stairs, and sleeping on four-inch foam pads, the team awoke in the morning to see their surroundings. It was, maybe, what we would call a one-star hotel. They were not certain that the sheets had been cleaned and were less than thrilled with their millet gruel for breakfast. This wasn't looking good.

Journey for One

Once Joan had taken her quick ice-cold shower, she set out to see the surroundings where she would work for the next 21 days. She was disappointed but not surprised to see that the facility looked more like a prison than a hotel and was actually the national headquarters for the meat, fish, and eggs industry. Needless to say, these weren't the warm and cozy surroundings that most American therapists had always worked in. They did understand though that this was a disaster situation and everyone was making do the best they could with what there was to work with.

The team worked beside local therapists and doctors to service those affected by the situation. They did the very best they could with the meager conditions they encountered. Joan had done work like this before. She knew it was not uncommon for people visiting locations such as this to feel uncomfortable and unhappy. The living conditions were less than ideal and could only be described as minimally acceptable at best. After two meals a day of cabbage and potatoes for the first three days, the team knew this wasn't going to be easy. They had choices to make. They could just not eat, probably not a healthy or reasonable option; or they could go ahead and eat the meals and try to keep criticism and dread at a minimum. Their beds were not comfortable but they chose to focus on their efforts to be of service instead of bemoaning a situation that was clearly not going to change. What the team of 12 decided, both individually and as a group, was to find ways to entertain themselves in positive ways that allowed them to continue despite the hardships. Some team members started a dance time every evening. Others began sketching, doing origami, and playing word games. They capitalized on their strengths instead of focusing on their challenges.

Joan explains that leaving the site and returning home was an almost impossible option. The airport had very limited traffic going in and out, was a two-hour car ride away, and there were very limited possibilities for transportation. Being flexible and making do became a way of life. Once they accepted this they noticed that despite the obvious downfalls, they were also encountering some unexpected benefits. They were making new friends with the local counseling staff, and they were able to visit a local school and play soccer with the kids. They spent time learning therapy techniques from the Russian psychologists, and when they had a day off, they were able to visit a town about an hour away and engage in some enjoyable time doing things they wanted to do. Joan's face lit up when she shared a story with me about meeting a group of gentlemen in a hotel lobby who were national managers of McDonald's restaurants, which had just opened in Russia. The gentlemen offered to buy the team lunch and chatted with them about life in Russia. Joan's favorite, and perhaps only, souvenir from that trip was a McDonald's T-shirt in Russian that she had gotten from a young man who traded the T-shirt for a pair of blue jeans Joan had brought from the U.S.

It's been many years since Joan made that trip to Russia. It was a trip that she could easily remember as one of the most awful times of her life, but instead, when asked about her trip there she instantly lights up, smiles from ear to ear, and begins to tell you of the many difficult things that were overcome on that journey and the tremendous joy she experienced through being flexible and opening her mind and her heart to possibilities. "I still have the T-shirt," she says with clear joy in her eyes. I asked her if she would do it again. "Oh my, yes," she crows. "No question. The next time though, I would explore more options and possibilities for travel in addition to the work we were doing. I'd be more adventurous and flexible. It would be fun to see what all I could learn."

Pack Your Bag

Life is full of ups and downs. We have situations in life that turn out better than we planned and others that go terribly wrong and disappoint us greatly. How we handle those disappointments and what we do with them is what is important.

As we begin to talk about things going wrong and what to do when they do, I'd like you to think about a situation you have had in your life when plans you made didn't work out the way you had hoped. Perhaps it was travel plans, maybe a birthday party plan, or even a work-related event that didn't turn out well. As you think of an event, think of one that you feel like you handled well and that turned out well in the end. Focus on a situation that lived up to the old Shakespeare play, "All's Well That Ends Well."

Briefly describe the situation that didn't go as planned, including what was meant to happen.

Now describe what it was you did to bring the situation to a happy ending.

Learning the Way

"Those who fail to plan, plan to fail." This bit of wisdom is attributed to many people. For the most part I believe it is true. It's a basic premise to my approach to solo travel. However, I also believe that no matter how well you plan, things sometimes fail anyway. We are not always in control of how things work out. Especially when traveling, many of the people, circumstances, and conditions we encounter are completely out of our control and can cause our best planning to go wrong. The secret, I believe, to not only surviving but thriving in difficult travel situations is our ability to remain creative and flexible in the face of that difficulty.

I have had many, many situations in my years as a solo traveler that have caused me to change my plans and have caused me various degrees of anxiety, ranging from minor annoyance to absolute panic. There are several tips I have learned from these experiences, which I believe make all the difference when handling a stressful travel event successfully.

- I can't always be in control.
- Being well prepared never hurts.
- Schedule myself loosely enough that travel delays have minimal effect.
- Remember to stay calm.
- Prepare financially ahead of time to take care of a travel problem.
- Maintain my dignity and treat others respectfully.
- Many of the unforeseen situations I have encountered while traveling have turned out to be my best memories.

Let's take a look at each of these seven tips and how they can serve you when things go very wrong.

I can't always be in control.
I like to feel competent. As a teacher for many years, I knew that the best way for me to do my job well was for me to be in control of my classroom as much as possible. I am also the oldest of three children in my family, which psychology says means I like to be in control. Add these factors together and you get a woman who is much more comfortable if she is in complete control of every situation.

When I started traveling solo many years ago, one of my first challenges was to accept the fact that sometimes you just have to give up control and let things be what they are. You can't control airline schedules, you can't control the weather, and you can't control the culture of other people and places. As I looked at why I was traveling to foreign lands, I remembered that it was to learn about new countries, ways of life, and to experience diversity in its native setting.

It took a few trying situations and a couple of pretty embarrassing failures for me to finally get the message that sometimes I just have to let go of needing to be in control and that I need to make decisions and choices based upon the reality and situations of where I am. This hasn't always been easy and I haven't always enjoyed these experiences, but they have certainly taught me the importance of calming down and really listening to what is happening. I have learned that the energy and emotion I spend on trying to fight a system I have no control over is a giant waste of time.

I do and will continue to question policies, schedules, and answers I get when trying to solve problems because you never know when you might be able to change something. But ultimately it's like my grandpa used to say, "Poking a bull with a stick is a waste of time. All it does is wear you out and pisses off the bull." Don't you just love grandpa wisdom?

Being well prepared never hurts.
In several ways, throughout this book, I talk about the importance of being prepared, planning, and thinking ahead. Preparation can mean the difference between success and failure. It is difficult, I know, to prepare for something that you've never done before. However, just because you may be traveling alone does not mean that you cannot prepare. Use every resource you have at your disposal to help you. Enlist your friends and family who are well traveled. The internet is a treasure chest of information. Blogs, websites and social media are like having travel experts in your back pocket. These resources can and will help you be proactive and ready to make the best of your travels. I have listed many of them in the Resource section at the end of the book. It's like when you take your swimming suit with you on a summer road trip to see your grandma. You may not need it, but how great is it if you happen to stop at a hotel with a pool, to have a suit and be able to swim blissfully after a long day on the road?

Schedule myself loosely enough that travel delays have minimal effect.
So many of us have a much bigger travel appetite than what our actual schedule and wallet can handle. We forget that it takes time to drive to the airport, wait in line to check in, shuffle through security, and deal with flights that get stuck at the gate for maintenance issues. We tend to underestimate how long everything takes and have very little control over much of what takes up our travel time. I'm one of those people who would rather show up at the airport earlier than necessary, then sit in a cafe and enjoy a cup of coffee, than to spend the precious time before my flight running and pushing my way through lines and crowds. If I schedule in extra time as I plan my travels, I can be more relaxed and have the time to do very important tasks like go to the bathroom before a flight. Details like this

cannot be underrated when we travel solo. Extra time can also make the difference for you in being able to see something you're interested in and being turned away because you won't make it through the line by the time a site or tour closes.

Remember to stay calm.
Trust me, I know how truly difficult this can be. When you have spent a lot of money, traveled for a long time, are hungry, nervous, and a bit anxious, it is easy to become frazzled and lose patience with any glitches that occur. Science has proven that our brains, under stress, do not perform as well as when we are calm. If you can possibly maintain your composure while all others around you are losing theirs, you have a huge advantage when you need to problem solve or be creative.

I know, firsthand, how it feels to be moving along smoothly and then some glitch happens and all of a sudden you feel like your whole world is falling apart. This can happen whether it's a small glitch or a more serious issue that comes out of nowhere.

Prepare financially ahead of time to take care of a travel problem.
Over the years I have had several young friends who have felt a sense of wanderlust and headed out for journeys with nothing more than a backpack full of hopes and dreams. They have their very limited stash of cash in their pocket and are certain they can make it last. Having had a little more life experience than they have, I can't help but put on my teacher hat and ask them if they are prepared for an emergency. Do they have enough money stashed away somewhere in a sock, a bank account, or a rich relative, that should they find themselves without a cent and trouble arises, they will have the means to take care of things financially? Often, the answer is no. They are optimistic and full of hope and are sure that nothing bad will happen. For those of us with a few years on our resume, we appreciate the optimism but know the reality of needing to prepare ahead of time.

When I talk about preparing financially I am speaking about cash in hand, but I am also talking about the ability to get ahold of funds quickly in case of an emergency. It might be a credit card that you keep just in case of such instances. It might be a friend or family member who has access to funds that can reach you in a hurry, or it could be travel insurance. I have been around the block enough times to know that although we don't often need to use the insurance we buy (thank goodness), it only takes once to need the insurance and have it available for us to realize what a wise investment it is. Regardless of what form of emergency funds you decide to arrange, I strongly encourage you to make sure you have a source of extra funds. It is one of the most important ways to feel more independent, capable, and self-assured when traveling on your own.

Maintain my dignity and treat others respectfully.

I once got grounded in Miami due to hurricane warnings in the area. There was no way that any plane going anywhere was getting off the ground. It doesn't take much imagination to picture a huge international airport full of travelers on their summer vacations, lined up in mobs at customer service counters. Everyone was trying to get the very next flight, the very best hotel room, and the fastest service for themselves. It was a nightmare. I happened to look up at one point and catch the eye of one of the service representatives doing her best to take care of a really bad situation. You could tell that she was just as frustrated, anxious, and tired as every single one of the travelers she was trying to help. I then looked at the reactions of the passengers around me. Some were taking the whole thing in stride, others looked scared out of their minds, still others acted as if they were royalty and deserved to be treated with special care and immediately. It occurred to me that we were all just human beings doing the best we could in a bad situation. None of us, including the customer service rep, had wanted to be here, deserved to be here, or were enjoying the experience. As I got to the front of the line and greeted my eye contact buddy at the desk, I realized that this lady could probably use a smile and a friendly word. After a few seconds of telling her how badly I felt for her in this situation, giving her a genuine smile, and relating to her that I was sure she would do the best for me that she could, this kind woman starting pushing buttons on her keyboard, checking various screens, and within a few seconds announced that she had just found me a room at her favorite airport hotel and a seat on the very first flight to Denver the next day. I didn't ask for any of that. I merely kept my cool, took the time to smile, and treated the poor woman with the dignity she deserved in this awful situation. It always touches me when I see someone's face light up because of a few kind words or a simple smile. Those two tools — a kind word and a simple smile — are the most valuable things you can carry with you when you travel.

Many of the unforeseen situations I have encountered while traveling have turned out to be my best memories.
Several years ago, while I was on a volunteer trip to Guatemala, one of the doctors and a couple of other staff from the clinic were going to a village about two hours away to offer a day of clinic services for children. Knowing my love of kids and my ability to entertain groups of children, the doctor asked if I would like to go along to help. I think I was in the van before he even got the invitation out of his mouth. There had been some civil demonstrations over the past month or so and we had heard rumors that the farmers were planning a protest but we had no further information so we proceeded with our trip to the village. On the way, a group of very creative farmers, instead of yelling and carrying protest signs, simply used their oxen to pull giant fallen trees across the road and then "accidentally" left them lying across the road so no traffic could move in any of four directions. I thought it was a brilliant strategy until we realized that the tree was going nowhere and therefore neither were we. The team and I chatted for a while, listened to the radio a while, and somehow got started on an English lesson for the team. Some of the group had a little bit of English and we ended up on the topic of the difference in the pronunciation between /sh/ and /ch/. We practiced on words such as sheep, chip, ship, and sheet. As soon as I used the word sheet we all broke out in hysterical laughter and I then had to spend about 10 minutes with them practicing the difference in pronunciation between sheet and shit. It was glorious! Here we sat, stuck on a highway, not able to move or get to our patients in the village for who knows how long and we were having more fun than we ever imagined. That morning in that van is one of my all-time favorite travel memories. What could have been a nightmare of a wait, with grumbling, complaining, and discomfort, turned out to be one of the best cultural experiences of my 47 years of travel. So, don't forget that sometimes if we can't change a difficult situation, it's better to relax and wait. It's entirely possible that something even better and more memorable is going to happen.

Let's now take a look at how, when these difficult moments during our travels occur, we can step back, get our "sheet" together and make the best of the situation.

I recently traveled back to Vienna for a visit. I scheduled one day pretty full of things to do, errands to run, and people to see. I knew exactly how I wanted the day to go so that I could do everything on my list. Well, wouldn't you know it, the first thing on the list, going to the bank, didn't work out as expected and suddenly the whole plan fell apart and so did I. I was angry, feeling helpless and frustrated, and wasn't sure whether to scream or cry. It took me a minute or so to remember that being in this state of mind was not helpful. If I was going to get past this and have a successful outcome, I was going to have to pull myself together. I sat myself down on a bench, took a few deep breaths, and thought, you can do this, Jodie. This is not the end of the world. We've all been there. It's a situation where normally it would not be a big deal, but when you are alone, traveling, and in unfamiliar territory, seemingly small things can feel like a big deal. Once I took that couple of minutes to calm down and get my wits about me, I was able to figure out a Plan B and move forward. It all worked out fine in the end and I went on to enjoy my busy day.

The same principle applies when what goes wrong is more serious. Whether it's a travel emergency, a serious medical issue, or you have been robbed, the very best tool you have is your ability to stay calm. I hope you never have an experience where you need to try this out, but you should feel better knowing that if you do run into difficulty, staying calm can make the difference between things falling apart or getting back on track.

So, you ask, how do I stay calm? The very first step in calming down is to stop. Stop walking, pacing, driving, or whatever you are doing and give yourself the gift of a little time. Until you are able to take yourself out of high gear and get closer to being in neutral, nothing is going to get better.

Once you are able to stop, take some deep breaths and keep reassuring yourself that you can deal with this, you are a competent adult, and no situation is permanent. This will give you valuable seconds, which will let your body and mind stop operating from adrenaline and start operating from a more relaxed and productive state.

Travel Triage

You are then going to go through the six steps of something I call Travel Triage. Triage is what they do in medical settings when many serious patients come in all at once. The purpose is to see who is in the worst condition, who can wait to be treated, and where patients need to go to get the best care. In my Travel Triage, I go through very similar steps. Let's take a look at the six steps and their purposes.

Step 1: **Describe**
This allows you to take a good look at the situation and see what the problem really is. You may discover that the problem really is not what you initially thought it was. Sometimes you may misidentify the problem and then go chasing the right solution to the wrong problem. I once spent over an hour and a half frantically pacing around a London airport looking for a tour company representative because she was not at the spot where I was told to meet her. It turned out, after I finally found her, that I had misunderstood the directions and was supposed to have waited by baggage claim. I could have saved myself a great deal of time and grief if I had just slowed down and really identified and solved the correct problem.

Step 2: **Calm Yourself**
This is the step I described above where you stop, take some breaths and reassure yourself that you can handle this. You can work through the difficulty and have it turn out well in the end. You will survive.

Step 3: **Remind Yourself**
This step is important to help you figure out what your immediate goal is and how urgent finding a solution is. Is the situation something that has to be handled immediately or is it something that you can take a bit of time to figure out? You don't want to make a snap decision that will create new and bigger problems later on.

Step 4: **List What You Can and Cannot Control**
There is absolutely no point in being upset about something you have absolutely no control over. Things like airline schedules, road closures, weather, and business hours are situations that you can't change. Yelling or demanding

things won't change these situations either. You are better served to focus on figuring solutions using the things you can control, like how to get somewhere, when it's best to go, and how much you are willing to spend to work out a solution.

Step 5: **Choose Your Best Two Options**
We've heard that it's always good to have a Plan B. You would, of course, prefer that your first choice is what you get, but in the event that this doesn't work out then you are way ahead of the game if you have a Plan B already figured out and ready to pull out at a moment's notice. So, as you are thinking of possible solutions, don't stop at one.

Step 6: **Take Action**
Occasionally when I am in a very tense situation my brain seems to think that sitting and doing nothing is a good idea. Clearly, that's not usually the case. However, taking a moment or two to stop and double check with yourself to be sure that what you are going to do still sounds right is not a bad choice. Once you've given your plan a chance to settle a bit in your head, then it's time to take action. I once heard a saying, "All planning ultimately degenerates into work." There comes that moment in our problem-solving that we just have to get up and make a move. Ignoring your travel problems will not magically make them disappear. Sorry!

I have created a chart below that shows you how this process might look in a real travel situation. See how each step helps put you into a place of control and creativity, which allows you to problem solve much more effectively and increases the chances of a satisfactory result. Again, you may not get what your plan was in the beginning, but keep the ultimate goal in mind. In my example, I wanted to get to the wedding and the new plan got me there.

It will be helpful to you to practice using this six-step process in your daily life prior to traveling solo. As my grandpa used to say, "It's tough to think clearly when you are ass deep in alligators!" Don't wait until you are stranded at the airport to try and figure out how to use this system for solving your problem.

Steps to take for Travel Triage	My information
Describe and clarify what the problem really is.	I got to the airport and found out that I had been bumped from my flight because of over selling of seats.
Calm yourself. Remember to breathe. Repeat a phrase that helps calm you down.	I need to stop and sit down. Take some breaths. Get something to drink. Continue to breathe until I feel myself calming down.
Remind yourself of your immediate goal and need. Decide if the difficulty is urgent or if you have some time to work things out.	I need to get to Montreal by 1:00 tomorrow for the wedding
List what you can and cannot control.	Can control: ○ Price I pay ○ How I get there ○ When I go Can't control: ○ Airline schedule ○ Delays ○ Weather ○ Airline policies
Choose your best two options from the things you can control, then choose the one that suits you and the situation best.	1. Fly on another airline on later flight. 2. Pay for upgrade to business class so there is an open seat. Deal with getting a refund from airlines or travel insurance later.
Calmly **take action** on making your first and if needed, second option happen.	Go to airline desk and make arrangements for the new ticket.

Adventure

Often, what feels like disaster in the moment, turns into our funniest travel stories after it's over and we are safely back at home. We never go home and report that all of our flights were one time, the food was perfect, the beds were all to our liking, or that we never had to wait in line. If you listen to someone tell of their travel adventures, their stories are often sprinkled with details of mishaps and mix-ups and how they got themselves out of the mess. These travelers will also frequently laugh or make fun of the situation, although I'm certain it probably wasn't funny at the time. I'm certainly no different. Some of my favorite travel stories are the ones where I got lost or things didn't work out at all like I had planned. Sometimes the outcome is even better.

This adventure involves you going out and finding people with travel stories to tell. I want you to ask specifically about a time when someone traveled, preferably alone, and had something go wrong. Listen carefully as they tell the following parts:

- What happened?
- What was the effect?
- How did they deal with it?
- How did it turn out in the end?

Listen to how they talk about it. Ask questions, if you like, to get them to tell you more. Pay attention to their general feeling about the incident. The more stories you can hear from a variety of people, the better you will understand that we all face challenges as we travel. It's not so much about what the situation was, it's more about what you do with the situation that matters.

Looking in the Rearview Mirror

You have now had a chance to talk with fellow travelers about their challenging travel experiences and heard, as the movie title suggests, the good, the bad, and the ugly. Rather than focusing on the things that could go wrong, let's instead focus on how things can work out well in the end.

7: When Things Go Wrong

Use the chart below to take notice of how the people you spoke to handled their difficulties. Then think of a time when you had a struggle while traveling, even if you weren't alone, and chart that to see how you handled things.

Name of Traveler	Person #1 Name	Person #2 Name	Me
What happened? *I missed my train.*			
What was the effect? *I would have to spend the night in the little village.*			
How did they deal with it? *I got a room at a little inn.*			
How did it turn out in the end? *We stayed up late into the night and played folk music and drank shnaaps.*			

Celebration

The whole idea of travel is to experience things that are new and different and that give us some escape from our normal and sometimes routine lives. Many first-time travelers list being afraid of trying new foods or not liking different foods as one of their top concerns when they travel. For some, this seems silly because they love all kinds of foods from around the globe. To many, it is a matter of never having tried unfamiliar ethnic dishes.

Your celebration for this chapter is to find a type of food that you may not have ever tried before. Look around your community or area and find a place that gives you the chance to try something new. It may be a restaurant, ethnic market, or even trying a recipe in your own kitchen. The internet is a great resource for finding such things and you might even find some surprises. Try something new, celebrate your latest step in this journey, and don't forget to have fun doing it.

8

STAYING SAFE

A ship in harbor is safe, but that's not what ships are built for.
—John A. Shedd

Travel Tale

For many years I traveled to Central and South America to work with medical teams providing healthcare to rural indigenous people. After spending a couple of weeks with a team of people 24/7 I was always in need of some *me* time. I would always schedule some time to travel on my own after each of those medical trips. After one such trip I decided to explore a beautiful Central American city in the mountains. I was excited to get out and experience the city since we had been mostly in rural areas.

One day I decided to visit a popular public market that was famous for handcrafted items and colorful food displays. The guide books all said it was incredibly crowded and to be cautious of pickpockets. I was carrying a shopping bag with my glasses in their case over my shoulder. I got so excited looking at all the people, handicrafts, and the bustle of the market that I wasn't paying attention enough to realize that the man who kept bumping into me from behind was actually a thief. One of the stall keepers at the market yelled to me that I had been robbed. I couldn't imagine how. I had intentionally carried a small wallet in my front pocket just like they tell you to. I didn't have on any expensive jewelry and wasn't flashing any money. How in the world ...? Then I realized that the bag had been cut with one swift slice and my brand new, expensive, prescription sunglasses were gone.

I felt every emotion you can imagine. I was frightened, grateful I hadn't been hurt, angry, scared, tickled that the thief had only gotten glasses they couldn't use, and thrilled that despite his best efforts the thief had been foiled. The lovely stall keeper made sure I was all right and then explained to me that the thieves were plentiful there and the problem was that the country had so much poverty that people resorted to any means necessary to feed their families. I thanked her for her reassurance, threw the shopping bag in the trash and continued on my shopping trip much more alert to my surroundings. After the initial shock wore off I became angry. How could they allow such poverty to exist? Why didn't they have more police presence at the market? Why would anyone ever want to come here for a vacation? Later that day I recovered enough to understand that what they say is true, "Shit happens."

Despite being what I felt was well prepared, I became a crime victim. In that place at that time there was nothing that could be done. The police would and could do nothing and there had been no real harm done. It was just something that can happen anywhere. It can happen in your hometown, it can happen in a large crowd at a sporting event, and it can happen on public transportation. The only thing we can do is be aware and prepare.

Be aware and prepare. That became my safety mantra from that moment on. I realized that each and every one of us has a different level of fear about things that could happen. We all have different ways to recover from bad things that happen. And we all react differently to challenges. I have learned to be as aware as I possibly can to the safety conditions around me and prepare by reading, asking questions, and protecting myself and my belongings in the best way I can. I am happy to say that I have never been hurt, robbed, scammed, or frightened seriously. I know others have not had the same good fortune. I also know that just as it is with catching a cold, you can be informed, prepared, and do your very best to prevent getting the cold, but sometimes despite your best efforts it happens to you. It's best if you take good care of yourself, get any help you might need, and recover in the best way you possibly can. Just like with that nasty old cold, you can't let one bad experience prevent you from doing what you enjoy forever.

There are basic things that I can do to keep from having problems. If I heed those basic safety rules on my travels, 99.9 percent of the time I'll be just fine. There's a whole big beautiful world out there, and I refuse to sit in my living room and play the "I Wish I Could" game. I want to travel and embrace the world.

Pack Your Bag

Thinking about bad things that happen isn't something you want to do, and it certainly does not help when you are already fearful about traveling alone, but being well prepared and aware ahead of time can make all the difference.

Think about all the places you have been or that you wish you could someday visit. Consider what you know about the types of things that might be safety challenges in those places. Perhaps it's pickpocketing or taxi cab scams. Maybe it is the scariest thing you think could ever happen to you while traveling. Write a list of those things and any explanation for yourself that you think is important. I have given you a typical fear that travelers often have.

My Safety Concerns	Explanation if Needed
Someone will steal my passport.	*Can't get back home or stolen identity.*

Learning the Way

Technically, I have been traveling on my own since the age of 14. My first solo journey was alone in the sense that I was without my family. I was with a school group on a trip to Mexico with my Spanish teacher. Once we got to our destination, we were taken to homes where we stayed alone with a local family for two weeks in order to be immersed in the culture and language. It was up to me, with the help of my host family, to learn how to navigate the city, be a part of family life, and participate in cultural classes set up by our teacher. To say I was scared would be a gross understatement. I was terrified! I wasn't sure my Spanish was good enough to get me what I needed. What if this family I was to live with wasn't nice? What if something bad happened? What would I do? Remember, this was long before the invention of the cell phone. I later found out that my other classmates on the trip were also scared and that my parents, too, were concerned, but we all trusted the teacher completely and it turned out that we were justified in our trust. He made sure, once we were there, to support us, have contingency plans for everything, give us guidance and reassure us that all would be well. I'm certain that our wonderful teacher felt tremendous responsibility for our safety. I realized later that he had made this trip many times with many students and that he had considered every possibility of difficulty and planned ahead, so that he could teach us to be cautious and safe on our own. He wanted to empower us so that we would have the courage and confidence to face difficulties in life, not just on this particular trip but for the rest of our lives. The trip turned out well and I learned so much more than I signed up for. It was truly a high point in my education and an experience that shaped the way I travel to this day.

I made this summer trip to Mexico every year for several years. I loved it. I learned to be comfortable, curious, and cautious. Thank goodness I had made all of these trips, because when I was older and decided to go solo I knew what was possible. I knew that, while it might be frightening and intimidating, I could do it.

When we travel in our own area or country, we are probably pretty familiar with the expectations, quirks, and problems that the area has. We know what neighborhoods to avoid and what we need to do, or not do, to stay safe. We know the rules, how to involve the police if necessary, and what to watch out for. When we are traveling alone we just need to turn it up a notch, be a little more aware of our surroundings and our behavior. Being in a place that we don't know and wanting to explore is great fun. What sometimes happens is that we get so engrossed in seeing and exploring that we accidentally forget to keep ourselves alert to possible dangers or what may be happening around us. I'm not saying we have to forget sightseeing and stay alert at every moment, I'm saying that by taking a few precautions, the possibility of safety issues is minimized as much as possible. It's about planning and setting things up ahead of time so that bad things are less likely to happen.

Let me illustrate this idea with a little story from my work life with university students preparing to become teachers. I was teaching a lesson about making rules for kids. I gave each of them a little container of modeling clay and told them to play with the clay and write down *all* of the possible things that a young child might do with clay that you wouldn't want them to do. They came up with putting it their nose, throwing it across the room, stealing it, putting it inside your clothes, eating it, and many, many other creative ideas. After we had shared all of the possibilities I said, "Now, I want you to group those possibilities together in categories and write a few rules that you can talk about before your students get their clay that will prevent most of them from doing those interesting things." My students did a great job of coming up with those precautionary rules and I'm certain that as they worked with children later, many great clay disasters were prevented. My point with this story is to demonstrate how working through a possible scenario and being aware of the options you have to prevent things from happening is very helpful.

Think about this... you are going to a beautiful sandy beach on your own. You have your towel, sunglasses, phone, and a bit of cash in case you want a snack or something to drink. The water looks so inviting that you decide to take a dip in the waves. What do you do with your belongings while you are in the water? There are very simple and creative options, but while you're standing there with the sand between your toes your options seem limited. If you just take a few moments in your room before heading to the beach, you can figure out a solution and be ready to go. Does this mean you will figure out every eventuality and prevent anything from ever happening? Of course not. But you can be prepared for quite a bit, and being prepared is a big part of being safe.

My safety when I was preparing to travel to India was the one thing I was a bit nervous about. It was a very different culture from my own and I didn't want to find myself in a difficult situation because I didn't understand what was going on around me. When traveling internationally, the culture of a country in regards to travelers, especially women, can be different each place you go. It is important to know what those cultural norms are. The role of women and the rules about women can sometimes be quite different than what you may be used to. In Middle Eastern countries, for example, women are expected to act in a certain way and the cultural rules around women's interactions with men who are not family are particular to their part of the world. The tolerance for outsiders not following these norms can vary from place to place, but it is good to be aware of such things to save you difficulties while traveling. These norms and cultural customs are not difficult to find online and most reputable travel guide books specifically point these things out in their descriptions of varying countries. So, when traveling to a location that may be culturally, politically, religiously, or otherwise different from your own I highly encourage you to do your homework. Research the cultural norms and safety information for that place and then use that information to enrich your journey. Not only will you feel safer but you will also see that destination through a much clearer set of eyes than you might have otherwise.

Don't let stories or other people's fear and dread prevent you from going somewhere. However, do consider the concerns associated with a particular location. Planning ahead makes you feel more confident, which in turn makes you appear more confident to others and therefore less likely to be a target of wrongdoing.

I would be remiss if I didn't include a note about heeding warnings from reliable and informed sources. There is only one instance in my traveling life when I was scheduled to take a trip in another country and canceled it because of concerns for my safety and well-being. It was a trip to the Middle East that I had long wanted to make. Just a few short months prior to the trip, political problems began. After a few weeks things in that country had escalated to the point that the U.S. Department of State issued a travel warning, much more serious than a travel advisory, for that country. It was not forbidden by my government for me to travel there, but they strongly suggested that Americans avoid the area. I decided to follow their advice and am glad that I did. It turned out that nothing bad happened in my destination at the time when I would have been there, but you never know. (See U.S. Department of State website information in the Resources section at the end of this book.)

8: Staying Safe

Below is a list of suggestions that I have collected, which have served me well over the years. There are times when I need to have additional guidelines for myself based on a specific location, but for the most part this list has, I believe, kept me safer, allowed me to be more confident, and therefore let me enjoy my travels more.

Safety Tips
Safety tips to help you feel more confident and less fearful when traveling:

- Don't completely avoid interaction with strangers. They can turn out to be wonderful sources of information and the vast majority are harmless. Just be cautious and don't leave your well-being in someone else's hands.
- Pay attention to your surroundings. Wandering freely in a new place is fun and exciting but make sure you keep one eye and one ear focused on where you are and what is going on around you. Don't be caught by surprise.
- Don't bury your nose in a map. Standing on a street corner looking very lost and confused with a camera around your neck *SCREAMS* "tourist!" You might as well paint a giant target on your forehead. Try to look at maps, etc. from a more inconspicuous place such as a cafe, restaurant, or shop.
- Figure out where you're going ahead of time. If you are headed out to a specific location or just wandering around a certain area, plan ahead. Look at a map and be fairly familiar with the layout or directions for a location. This goes along with not walking around with your nose in a map.
- Don't flash the cash. It's tempting to walk away from the cashier who has just given you a pile of bills as your change and put them away in your wallet as you continue walking away. Don't do it! Take the few seconds you need to stand at the register and put your money away. Also, don't pull your cash out of your wallet while in public and count it. You might as well say "Hey, look what I have! Do you want some?"
- Stay off of your phone. I know it's popular today to spend time while walking around chatting with your friends on the phone. This practice takes away the attention you should be paying on where you are and what is going on around you.

- Be cautious with whom you choose to sightsee/travel. I know it is tempting when you are alone and wishing you had someone to join you on your outings to befriend a stranger and join up together for excursions. This is fine. I've met some wonderful people this way. Just spend some extra time and be very picky about whom you decide to befriend.
- Observe precautions the same as everywhere: avoid being alone at night in dark places. At night stay in populated and well-lit areas as much as possible.
- Let at least one person know your itinerary daily via text message, email, phone, etc.
- Stay aware. Pay attention to what is going on around you. It's a tragedy that our world has gotten to the point where we have to be on our toes at all times to protect ourselves, but it's true. Chances are that nothing will happen but, as the old saying goes, "Better safe than sorry."
- Do your homework by researching your destination thoroughly. With as many wonderful resources as we have available on the internet and in print, it adds even more fun to the trip by being able to get excited ahead of time by learning little, or big, tidbits of information about your destination. Know before you go is good advice.
- Have extra documents separate from the originals along with you. Make two copies of your passport, visa, credit cards, immunization records, etc. before you leave home. Carry one set with you in a different location than where you are carrying the originals. Perhaps put the originals in your bag or backpack and then safely store the copies in your suitcase and/or the room safe if there is one. Give the second set of document copies to a trusted friend or loved one so that in the event that you lose your originals you have a way to get the information quickly.
- Trust your gut. They say that if something seems too good to be true then it probably is. If something doesn't seem right and it makes the hairs on the back of your neck stand up, there is probably a good reason. You can walk away, find someone for help, or make an alternative plan. If it feels shady it probably is.
- Use known and reputable tour/guide services. This is not only to avoid shady schemes but is also important for insurance purposes and legal rights in the event of an accident or problem.

• 8: Staying Safe •

- Ask at hotels, tourist offices, police, etc. about safety at a given location if you're not sure. The tourist industry in any given location is important to the local people and businesses. Their success depends on your well-being and happiness. They are more than happy for you to ask about an area or location and whether or not it is safe for you to be going alone.
- Leave a note in your room saying where you're going, when you left, what you are wearing, and perhaps when you will be back. In the event that there is an accident or problem this information would be very helpful to the authorities in getting you help as quickly as possible.
- Carry an extra $50-$100 hidden away in your purse, wallet, bag, or backpack in case of an emergency. This amount will most assuredly get you somewhere warm, safe, and comfortable should you get lost, sick, or afraid.
- Have a retreat plan. There is nothing in the traveling solo rulebook that says you have to finish every activity you begin. If you find that you are not enjoying or not feeling comfortable with where you are or what you are doing you can certainly leave. I suggest making this plan ahead of setting out for the day. It's like the old Boy Scout motto: Be prepared!
- Make your personal safety and well-being your first and foremost priority. I haven't been on an adventure yet that was worth losing my life or well-being for.
- Be aware of mainstream media stories but check out the details elsewhere. There are trouble spots all over the world. But for those of us who have traveled quite a bit we know that what appears to be a big story implying that a problem is a national emergency often turns out to be an isolated incident. This causes unnecessary fear for both you and your loved ones at home. I once read a story about a woman who announced that she was canceling her plans to travel to Canada because there were shootings in the streets all over Canada and it wasn't safe to travel there. She had listened to a television news report about a shooting in Canada. The story had been sensationalized and it sounded like there was an all-out civil war breaking out. Once someone helped her research a bit online she was able to reassure

herself and her loved ones that this was an isolated incident and did not mean she needed to cancel her trip.

- Be informed. As we know, the news is filled with stories that we later find have been sensationalized or exaggerated for the purpose of gaining readers or viewers. News sells. It's a fact. Responsible journalists do the very best that they can to accurately report happenings around the world but their reporting is only as good as their sources. My best advice is to never depend solely on one source of information in regards to possible difficulties. A true problem will be reported in more than one source and you will begin to see similar reports of the events and conditions. I believe it is also a good idea to limit the amount and source of your news gathering when preparing for a trip. Sometimes too much information is not a good thing. Stick with a few reliable and respected sources.
- Check out the U.S. Department of State Travel Advisory website (even if you're not American) or similar websites for up-to-date and accurate information on any country in the world. Any risks or cautions will be explained and suggestions made to help keep you safe.
- Stay aware and well informed. Sad but true, we live in a world that can sometimes change in a way that nobody wants.
- Understand the source of the warnings you are getting. Are they coming from knowledgeable and reliable sources or are they coming from fearful non-travelers, well-meaning loved ones, or sensationalized media reports?
- Dress and act sensibly. We are at a time when women should be able to travel freely and be entirely safe. Unfortunately, this is not the case. There is data that shows that especially when traveling alone in foreign countries, it is safer for females to be careful about how they dress and how much alcohol they drink. These two factors have proven to attract unwanted attention in some locations. Again, subtle cultural norms may be misunderstood and could lead to women unknowingly getting themselves into dangerous situations. This is a sad and unfair situation, but worth your attention.
- Stay comfortable, curious, and cautious. It will make a world of difference in keeping you safe and happy on your solo travels.

• 8: Staying Safe •

Adventure

Now that you can see how important and helpful planning for what *might* happen can be, take a look back at your list of possible safety concerns from the Pack Your Bag section. Using the location of your Dream Destination, take each item that was a possible concern for your safety and explore ways that you might protect yourself and prevent something from happening. I have, as usual, given you an example.

Dream Destination:

Safety Concern	Possible Preventative Measures
I will be going to a concert at night in a different part of town than my hotel. It will be dark, an unfamiliar location, and in an area with lots of bars and night clubs. I'm worried about my personal safety and how to get around alone after the concert.	○ Take a prescheduled taxi to and from the concert. ○ Ask the hotel staff about the safety of the area and get suggestions. ○ Look into group tour to the concert. ○ Wear more conservative clothes. ○ Stay sober so I can react quickly if needed. ○ Research crime rate of that neighborhood online. ○ Explore transportation options. ○ Maybe carry a personal safety device (pepper spray, whistle).

Looking in the Rearview Mirror

We generally think of taking a trip as something we look forward to and don't want to think about the "bad stuff." Especially when we are already nervous about the idea of traveling alone. Thinking about possible dangers makes us feel even that much more vulnerable. Rather than feeling vulnerable, this chapter was designed to help you see how empowered and prepared you can be when you are on your own. Just as a tennis player wouldn't show up at the tennis court without a racket, tennis balls, some water, and a towel, as travelers we don't show up for a trip without some preventative measures, a keen eye, awareness, and a plan. It just wouldn't make sense to do it otherwise.

Now I'd like you to challenge yourself to go somewhere locally that you have never been before. Think of a place that is relatively safe because I'd like you to go alone. Think about what you've learned about safety and think through two or three things that might be safety issues. What could you do to prepare for your safety in that location?

After thinking about your safety precautions, go ahead and visit the local spot that you planned for. Try out your safety plan. Did you feel more confident? Were you able to enjoy your visit there? Was there anything that surprised you? How did it feel to be prepared and empowered?

Celebration

You have just taken a huge step in empowering yourself as a solo traveler. What an accomplishment that is! You should be very proud of yourself.

This chapter's celebration is one for you to relax and enjoy! Find a cozy place to relax and watch a movie. Find a movie or video that is about or features a location that you've always dreamed of visiting. Perhaps it's the Greek Islands so you can watch *Mama Mia*. Or, if it's Antarctica you can watch *March of the Penguins*. You choose. Sit back, relax and think about how safe and competent you will be when you get there. Way to go!

9

HEALTH MATTERS

An ounce of prevention is worth a pound of cure.
—Benjamin Franklin

Travel Tale

I was raised by a nurse. My mom worked in emergency rooms for much of her early nursing career. Any of you who are from a similar background understand how that works. Unless there is a bone protruding from your skin, a fever over 120° F, or you have blood coming out of two or more places on your body, you don't stay home from school. OK, it really wasn't that bad, but you get the idea. Because of the influence of a good nurse in my life, I learned at an early age about preventative health, healthy living, and taking care of basic injuries and illnesses myself. As I began to travel alone, I realized this training would come in handy and save me a great deal of trouble.

    I had been working in Guatemala as a medical translator one summer in the '90s. I had been serving at a clinic that provided medical care to the poorest of the poor in rural areas. It was the morning I was to leave the community and head back to the capital city to rest for a few days and then head home. I had made a quick visit to the neighbors to say goodbye, and as

Journey for One

I was leaving I tripped and fell face first down a short flight of cement stairs. As if that wasn't bad enough, there was a motorcycle parked at the bottom of the stairs. I truly think I had made it down the stairs with no damage, but it was my forehead hitting the rim of the motorcycle wheel that caused the problem. I sat up, tried to remember to breathe, and then realized I had blood running down the side of my face. Wanting to stay composed and look in control, I got up, held my hand over my forehead, and headed back to my empty house next door. I looked in the mirror to see that I had a pretty good-sized gash on my forehead. I wasn't too worried because I knew that head wounds bleed a great deal even if the wound itself is not too big. I had never had any formal medical training but from working with the medical teams and growing up with a nurse I knew that pressure on a bleeding wound was the thing to do. I grabbed what was left of the paper towels, got them wet and applied pressure to my head. Each time I checked I was bleeding less and less. But each time I moved, the cut would start bleeding again.

I realized at some point that it might require stitches. I had some decisions to make. The van to take me to the capital was due in about 30 minutes, and I didn't want to miss that. There was no medical care available in the little village where I was except the local doctor who worked on our team, and he had gone away for the weekend. I knew that I would have to do the best I could until I got to the city. The trick was making the two-hour trip to the city without bleeding all over myself and the van.

I always pack an emergency first aid kit to take with me when I travel. I could do a lot with what little bit I had. I thoroughly washed the cut with soap and water, put a little antibiotic ointment on it, covered it with a clean gauze, then wrapped my head with an elastic bandage like you'd use for a sprained ankle to keep the whole thing in place. I grabbed some ice from the refrigerator and made myself an ice pack with a grocery bag. I knew all of this would keep me going until I could get to proper medical attention in the big city. The van driver arrived, looked at me with a rather terrified look on his face, and then helped me load my bag into the back of the van and away we went.

I can only imagine what the staff at the front desk of the hotel thought when I checked in. They asked me if I needed a doctor. I decided that the hotel doctor in this case was probably going to be perfectly adequate and agreed to have him see me in my room.

The doctor was a very kind and helpful man. He had, by coincidence, done a little bit of medical training in the U.S. and was happy to practice his English with me. After checking out my wound he cleaned it up, praised my improvised medical skills, used a couple of butterfly bandages to keep the wound closed and then went on his way. I don't remember now what the charge for his visit was. I remember thinking it was a bit high but was so thankful that I was going to be OK and that no further treatment was necessary. Whew!

I have to say it took some light painkillers and an evening of using an ice pack to get me back up and running at full speed, but I was certainly glad I had not had to resort to any other options to deal with my injury.

While this was not a significant injury, it still taught me a very good lesson about how, when I travel, I am my very own first medical responder. If something goes wrong, as it sometimes will, I want that first responder (me) to be as prepared as possible and take the best care of me I can. That, in a word, takes one thing: Preparation.

Believe me, if I can deal with this kind of medical situation, anyone can. I'm not one of those people who does well with blood, shots, etc. but you do what you have to do and if you just have a little faith in yourself, a good first aid kit, and determination, 99 percent of the time you've got everything you need.

Pack Your Bag

One of the biggest fears many first-time solo travelers express is being hurt or sick while traveling alone. Certainly, injuries and illness can happen when you travel. They can also happen when you are at home in your own house. So, what is it that makes us scared of such things when we are traveling?

What I'd like you to do is think about any concerns or fears you have about getting hurt or sick when you are traveling alone. Spend some time really thinking this through. Have you heard scary stories? Have you had something happen when you were traveling? Do you have a condition that makes it more likely that you will become hurt or sick? After you have thought about your concerns I want you to write them down to use later in the chapter. No concern or fear is too big, too small, or too silly. If it worries you, it counts.

Yep, I've given you an example!

What worries me...	Why it worries me...
I won't know how to say bandage in their language.	If I am hurt and bleeding and I can't say what it is I want, I won't be able to take care of my wound myself.

Learning the Way

I believe that if you listed out all of the concerns and questions that make people hesitate to travel alone, you would most likely see that they all end up being about vulnerability. Being vulnerable means we feel more at risk, unprotected, or endangered while traveling alone than those who travel in pairs or groups. Whether that is true or not, that's often how it feels. When we are sick or hurt we feel even more vulnerable and frightened. Let's see if we can't demystify some of the thoughts and fears about health matters when traveling on our own.

I was well into my growing-up years when I finally figured out that the Boy Scout motto "Be prepared" was applicable to more than just the boys and that it was valuable to more than just the Scouts. Pre-planning for things that we fear but that may never happen is a great way to diminish our fears and allow us energy for more enjoyable tasks when we travel. We never know when we might bump into something, fall, become an insect's lunch, or catch a flu bug. However, we can be ready for most health-related matters long before we arrive at our destination, giving us a sense of security as we start doing, seeing, and eating everything we want on vacation. Think of it in terms of your country's military. They don't wait until someone attacks before they train, build up their resources, make plans, or locate allies. That

9: Health Matters

would be ridiculous. What is it that Jack Dempsey supposedly said? The best defense is a good offense? Let's assume what he said is true and approach concerns about health from the same position.

In today's work world, there have been many laws and regulations for worksite health and safety. Some businesses are required to put up signs reminding employees to wash their hands after they use the restroom. Others require goggles or special suits or special training. They do this for a reason. The companies don't want their employees getting hurt or sick because it costs them money, time, and paperwork. Shouldn't we do the same for ourselves before and while we are traveling? We can't just *hope* that we don't get hurt or *cross our fingers* that we don't get sick. But we can prepare. Let's look at all of the things you might need to prepare for, health-wise, while traveling. You might...

- Fall down.
- Get traveler's diarrhea.
- Get a splinter.
- Get hit by something.
- Twist a joint/muscle.
- Pull a muscle.
- Cut yourself.
- Have an allergic reaction.
- Get a rash.
- Break a bone.
- Get bitten by an insect.
- Get food poisoning.
- Drink bad water.
- Catch a cold or the flu.
- Drink too much alcohol.
- Burn yourself.
- Get a fever.
- Have pain.
- Feel heartburn.
- Get sunburned.
- Have eye irritation.
- Experience constipation.

Oh, my gosh! That's a pretty scary list if you look at it all at once. But really, look at that list again and see how many of those things have happened to you in the last two weeks. Unless you have terrible luck, you probably didn't experience many, if any, of these conditions. Most likely you won't experience them while traveling either. Having said that though, it is important to prepare.

When we are on vacation we often participate in activities that are a little higher risk than we might at home. We eat food and drink things that we are not used to and we come in contact with viruses and germs that are not a part of our usual environment. One way to calm our nerves as we plan and to feel confident in our ability to cope, is to prepare, prepare, prepare.

You can be way ahead of the game if you do what the army, the Boy Scouts, and I do, and prepare for most any possibility. Let's look at those military preparation elements again as a solo traveler.

TRAIN

Before it's necessary and you are all alone, spend time learning first aid. You can take a class, read a book, learn online, or learn from a medical professional. In any case, knowing basic first aid will give you an advantage if something happens and will give you confidence in knowing that you know what to do should the need arise.

BUILD UP RESOURCES

Our military puts a great deal of thought and planning into the weapons and supplies that they purchase. They would not want to buy snow shovels and thermal underwear if they were going to be working in the tropics. In the same way, we wouldn't want to carry personal fans or anti-malaria medication if we were going to Antarctica to see the penguins. We do what makes sense after we have done some research about the location and conditions we are going to. One of the ways that many solo travelers make sure they are prepared for health emergencies when traveling is to put together a first aid/medical kit to take with them. They contain the essentials for wound care, basic treatment of health symptoms, and usually some printed instructions about how to deal with different things that may happen as a result of new and different travel experiences.

9: Health Matters

One of the most helpful and proactive things you can do as a solo traveler is to spend the time and money preparing a first aid kit to carry with you in your suitcase or backpack. It isn't necessary to carry a whole box of bandages or a full bottle of pain relief medicine. You need to carry only enough to get you through a day or two. If you aren't better by then with what you have tried, then it is probably time to find the services of someone with more medical training. Some of the things that have been suggested to carry on your travels are listed below. Some, like the elastic bandage I wrapped around my head, can have multiple uses.

- Cortisone cream
- Antibiotic ointment
- Allergy symptom relief tablets and ointment
- Acetaminophen/pain medication
- Bandages of various sizes
- Flexible bandage for stabilizing limbs or wrapping wounds (E.g. Ace bandage)
- Antacid tablets
- Antiseptic for cleaning wounds
- Gauze for wound cleaning and wound cover
- Bandage tape
- Small scissors
- Tweezers
- Butterfly bandages
- Anti-diarrheal medicine
- Emergency names and numbers, insurance info, home doctor's phone number
- Chamomile tea (or favorite tummy tea)
- Antihistamine/decongestant
- Cough drops/throat lozenges
- Professionally-made ice packs or zipper-style plastic bags to make ice packs
- Anti-gas medicine
- Nausea/upset stomach medicine
- Thermometer
- Eye drops
- Sunburn relief cream
- Water purifier
- Insect repellent
- Insect bite relief
- Laxative

You certainly do not need to carry every single one of the things on this list. It will depend on where you are going, what you are doing, and your own personal medical needs. Pre-made first aid kits are available for purchase; I bought one for $16 U.S. online. It had 299 pieces, and I added/subtracted some things to customize it for my own needs. It works just fine and was much cheaper than buying all of the contents individually. The kit is only 9 inches tall by 7 inches wide and 2 inches deep, very easily packed in a suitcase or backpack.

A kit like this has gotten many solo travelers through the bumps, bruises, and bugs that you normally encounter while traveling. Doesn't it feel better to know you could take care of yourself while traveling? It's like taking along your own little nurse when you travel.

MAKE PLANS

For those who have ever been a babysitter or hired a babysitter, you know that one of the most helpful things parents do when leaving you in charge is to tell you where they are going, how to get ahold of them, possible problems, and what to do in certain situations. Making a plan for your health and well-being needs before you travel is similar. It's like being your own parent and making sure that you have everything in place to ensure an enjoyable and less stressful trip. I'm not talking about planning to be sick or hurt, I'm talking about thinking through what to do in case you do get sick or hurt and the first aid kit you brought is not enough. It's for those times when you really need some help, someone to talk to about your situation, or when you are not able to take care of a problem on your own.

As scary as it seems, it's important to think through what kinds of situations might occur where you would need help. It might be that you've been sick for a couple of days and are getting worse instead of better despite doing your best with your first aid kit, staying in bed, and eating only tea and toast. It could be something such as falling down and knowing that you have broken a bone or that you have had strong pain for longer than you think is acceptable. In these cases, you may want to get more help. If you have taken the time to do a little research and planning at home before traveling, you will be ready for any such emergency and it will help you remain calm to know you are prepared.

9: Health Matters

Having the wisdom of someone who has done what you want to do and has gone where you want to go can be very helpful. You get the benefit of their wisdom, the lessons from their mistakes, and a shortcut to doing things the easy way. For that reason, I am going to share with you tips and recommendations that have served me well over the years and which will help you travel like a seasoned professional. To keep myself safe, in good health, and able to cope with any illnesses or injuries, I highly recommend that you prepare the following before your trip.

- Always check your regular health insurance prior to traveling. Many health insurance policies do not cover international travel or have strict policies about what they will and will not cover while you are traveling.
- Get extra travel insurance. Regular health insurance does not usually cover extraordinary costs such as a medical evacuation by plane in the event of serious illness or injury. While there is a cost to purchasing travel insurance, it covers a multitude of things from replacing luggage to trip cancellation and many items not generally covered by your personal insurance. It's worth checking into.
- Be prepared to adjust your expectations for emergency room, hospital, and clinic conditions when traveling abroad. We are very fortunate in the Western world to have top-notch medical care in beautiful facilities. Not all places have the same environment. The good news is that although the facilities themselves may be less modern than we are used to, the doctors at these facilities are often very capable and many have spent time training in other parts of the world and truly want to do the best they can for you.
- Remember if you are using a facility that is not up to your usual standards, to remember that it is not bad, just different. I have been treated by doctors on my travels that had none of the credentials that we associate with a high-quality physician, but I have gotten the best they had to offer and was far better off than if I had had nothing at all.
- You can always call your doctor at home. Be sure to carry your doctor's name and phone number with you. In the event that you are worried about the care you need or how you are being treated, you can always call your regular doctor. Often, if you tell the person who answers the

phone what is happening, they will do their best to get you the help you need as soon as possible.
- Pharmacies in many countries outside of the USA operate differently. Pharmacists in some places are allowed to give medical advice, give you medication without a prescription, and recommend treatments. If you find yourself in a situation of not feeling well, go to a pharmacy. Many international pharmacies will be able to give you medicine on the spot to help you. I found on one visit to Mexico that the pharmacist there knew the very best medication for my upset stomach because it was such a frequent problem for tourists. Asking a pharmacist is worth a try.
- Make sure that you have specific knowledge of possible concerns regarding where you are traveling. For example, if you are going to high altitudes, find out about altitude sickness and what you can do to prevent it. If you are going to a tropical location, you may need to take preventative medication for malaria. Numerous websites, including the U.S. Department of State and the Centers for Disease Control have information about health precautions for every area of the world. These sites can tell you what conditions you need to be aware of, any precautionary steps you should take, and any outbreaks or unique diseases that could affect your travels. Travel medicine clinics, websites, and doctors throughout the world are more than happy to help you prepare for a trip.
- Keep track of any medical treatments you get or medication you take while you are traveling. In the event that you need to see a doctor or get medication at a later date for the same condition, your doctor will want to know what was done and what medication you took to make a well-informed decision about your care.
- Remember to keep track of medical treatment and pharmacy receipts in case you are able to get reimbursed from insurance later on.
- Be sure to get an International Certificate of Vaccination, also known as the Yellow Card. It certifies the immunizations you have had so that your status is documented in case you travel to an area that requires a vaccination(s). These cards are generally available from your doctor or local health department. Travel medicine clinics should also have these cards available.

- Consider downloading the First Aid app from the American Red Cross at redcross.org/apps while you're still in the U.S., so you can access the content even if you don't have phone service. This app puts expert advice for everyday emergencies in your hand.
- Make sure to take all of your medications with you. Do not pack them in your regular suitcase if your bag will be checked in or separated from you during your travels. Also, be sure to carry a list of your medications, both prescription and over-the-counter, including the name, dosage in milligrams, and how often you take the medication. I always put one copy with my medication and one copy in my purse or bag. It is important to list all over-the-counter medications too. This includes vitamins, supplements, herbal remedies, and any non-prescription medications. It is sometimes important for medical personnel to know all that you are taking to check for any interactions or symptoms that may be affected by your over-the-counter medications.

LOCATE ALLIES

Prior to taking off for your wonderful vacation, take a few minutes to explore the medical facilities located near where you are traveling. Find out if there are hospitals, clinics, emergency facilities, or private doctors who service out-of-town visitors for medical assistance. Once while traveling in Istanbul, Turkey, my traveling companion lost her eyesight. We were scared to death and very concerned about the type of medical care she might receive. Little did we know that there was an American Hospital in Istanbul that took her in immediately. The doctor who saw her was the head of the department and was trained at Georgetown University Medical School in the U.S. I have since found out that there are many American and other nationality hospitals outside of the U.S. Many are set up specifically for the needs of ex-patriots living abroad and tourists. This information is very good to know before leaving home. Keep the name and contact information with you as you travel in the event you should need it.

The United States Department of State provides a service called the Smart Traveler Enrollment Program where you can register online as a traveler. You are then on a list of people who will be notified and located should there be a need while you are visiting. They are also there to help U.S. citizens in need of assistance during travels abroad. Other countries have similar services through their governments and embassies as well. Look into this prior to leaving home.

As you begin to plan for your health and well-being on a trip, there are many resources available to you. Here are a few of the sources I have used in the past for information related to my health questions.

I happen to be good at quite a few things, but I am not a trained medical professional and am not giving you medical advice here. If you have questions or are unsure about health-related information, please be sure and check with a qualified medical professional.

The International Red Cross is an agency that provides help to travelers worldwide.

Websites like preparedness.com recommend apps that inform you of where disasters are predicted and tell you what to do in the event of a disaster. The following free disaster applications install easily in your smartphone. While you may not use all 10 at once, you are certain to find three or four apps that you might want to use.

Tech Tip: Disaster Preparedness Apps
1. **Disaster Alert** by the Pacific Disaster Center provides mobile access to multi-hazard monitoring of and early warning for active hazards around the world.
2. **First Aid by American Red Cross** is one of my favorite apps and I highly recommend it. They also have Hurricane, Wildfire, Earthquake, and Tornado apps. The first aid app gives you simple, step-by-step instructions on what to do for any medical emergency. If you live in an area prone to one of these natural disasters, you should have the app!
3. **Global Emergency Overview** is an app that allows you to quickly browse through different countries and find out what is going on.
4. **Humanitarian Kiosk** was created by the United Nations and provides a range of up-to-the-minute humanitarian related emergencies from around the world.
5. **Real Time Warning** offers warnings about disasters around the world. You select an event to see its location, damage, severity and rumble radius on a world map. Especially handy of you have family and friends living in another country.

6. **Earthquake Alert** gives information <u>about earthquakes</u> with a magnitude of 1.0 and up in the USA, and a magnitude 4.0 anywhere else in the world.
7. **Siren GPS** lets you contact emergency services with the tap of a button. It will instantly give them your exact location and personal details. Wonderful for those that like to adventure in out of the way places.
8. When you push the **Red Panic Button**, this app will email or text your GPS coordinates and a link to Google Map to people you have previously set up. Like to hike or backpack in the wilderness? This is for you!
9. **Life 360** allows you and your family to set up a private network, then with the click of a button, you can let your family know where you are and if you're safe.
10. In the event of an emergency at work, **Guardly** enables you to receive emergency and operational alerts from your company or your security team. No more guessing.

Your very own home doctor is one of your best allies when you are considering travel. It is always a good idea to consult your doctor before traveling, especially traveling alone. Your doctor can tell you if it is safe for you to travel, any precautions you should take when you are away, and any other information related to your particular medical history that will be helpful to you as you travel.

Well, you now have a great deal of information to help you plan for and cope with any fears you may have regarding illness and injury while traveling. I want to repeat here that the secret, the absolute key to successful and healthy travel is being prepared. Don't be caught by surprise. Be your own best health ally.

Adventure

The lesson in this chapter gave you a great deal of information to think about and suggestions for things to do to prepare. It has ideas to help you alleviate most of your fears and concerns. Let's see if it took care of your particular worries.

Your adventure for this chapter is to think about your Dream Destination. Think about the fears that you wrote down at the beginning of the chapter regarding your health and well-being while traveling. Look and see if there are any fears or concerns that you had in the beginning that have not been addressed by the information in this chapter. If you still have concerns that were not addressed, do your research and see what you can find.

Now that you have looked back at all of your fears and the information I gave you, I want you to do the homework. Consider it an adventure into health and happiness. Do the research, download the app, look at the websites, write down the medical facilities, and make up your medication list.

You have a Dream Destination in mind. Make your homework and your planning fit your Dream Destination. Why not figure out what you would do if you got to have your dream?

Note:
The last two chapters have given you a lot of information about some pretty scary things that can happen when we travel. But, just like a wheel might fall off of your car, how many of us have actually had that happen? I'm certain it is very, very few. What I'm asking you to do here is to empower yourself with information, and that will go a long way in relieving your fears. You will get a chance in Chapter 14 to discover ways to look at your fears realistically and skills for being calm and confident even when you are afraid.

Looking in the Rearview Mirror

Find a friend or loved one who knows about and understands your interest in traveling solo. Sit down with them for a chat and share your homework from the Adventure section of this chapter. Ask them to help you think about whether or not you have thought of everything you might want or need to know. Add any suggestions or ideas to your list of information.

Celebration

Once you feel you have explored every concern regarding your health and well-being while traveling, do something healthy with your friend or loved one: Go out for a walk, go swimming, go get a smoothie together; anything healthy that will remind you that you are in control, you are capable of taking care of yourself, and that you can be healthy, happy, and whole while traveling on a Journey for One.

10

COMMUNICATING WITH HOME

If you have a choice of two things and can't decide, take both.

—Gregory Corso

Travel Tale

I had been invited by friends to come and visit them in the beautiful country of Montenegro. They explained that they would be working during the day and that I would be on my own. Great! I loved that plan. We would get up in the mornings and all get ready for the day. On their way to work my friends would drop me off in the city center or bus station so I could begin my adventures. One afternoon, several days into the trip, I happened to be feeling a little lonely and decided to treat myself to a nice lunch near the beach to cheer myself up. I discovered the most beautiful, peaceful and amazing restaurant in the whole country. I was sitting in the sunshine beneath a grapevine-covered pergola sipping wine and eating fish that had been caught not more than 20 minutes earlier. It was one of those perfect moments of life when you don't even want to move for fear of ending the bliss.

As I sat there I admit that a little bit of sadness crept in because of how nice it would have been to have someone there to experience and share it with me. Who would that be? I wondered. My mother, I thought, she appreciates moments and places like this. I decided the time difference was in my favor so I pulled out my cell phone and dialed her number. I didn't really care that it would cost me a pretty good amount. I just needed to share this with someone. Mom and I chatted for about five minutes and I explained the scene around me and the fabulous fish lunch. She sat quietly and listened and said all the right things. I thanked her for sharing that special moment with me and we said goodbye. For the rest of that trip I was happy and calm because I had been able to share a special moment and include that as part of the wonderful recollections of my trip to Montenegro.

As a side note, those phone calls have become a tradition between my mother and I. When I travel and am gone for more than a few days, once during the trip I will find a special place that I want to share with her and I will sit down for a coffee or a glass of wine and tell her about what I'm seeing. She still responds very positively and supportively and then we say our goodbyes and go on about our days. This practice has become a special routine for us and a ritual that I know I will treasure for many years to come.

Pack Your Bag

There are many reasons that you might want to contact a friend or loved one while traveling alone. There are also probably just as many reasons why you wouldn't want to contact home while you are away. Depending on your purpose for traveling, your location, the characteristics of the relationships, cost, specific situations that might need attention, or a number of other possible reasons, at one time or another we all struggle with the question of whether or not to contact people at home. Only you can decide what is right for you and your dear ones. Spend a few minutes brainstorming about positive reasons to communicate and problems with making contact while traveling. At this point I want you to make no judgments, don't think about what you think you *should* do, just think about what, in your heart of hearts, are your honest thoughts. Once you have done that, move on to the other sections of this chapter and we will revisit these lists a little later on.

Positive Reasons to Communicate	Problems with Communicating
I will get to tell my friends and family a little about what I'm doing.	I will have to hear about the problems that I am trying to get away from.

Learning the Way

After my first solo weekend getaway to the mountains before I went to St. Louis I was so pleased with myself and so excited to share my success and future plans. But then I realized I hadn't told anyone I was pursuing this endeavor, so I'd have to fill in the blanks before I could announce my great success. I knew this was an element of my traveling that I would have to give some serious thought to. My desire, style, and frequency of communication has changed over time.

It is important to know that much of my experience with communication happened before the invention of the cell phone, the internet, and simpler overseas phone calls. My approach to communicating with my family and friends evolved. It took a few trips before I came to what felt comfortable for me and did not feel intrusive, weak, or pitiful.

I now know that the choice of who to communicate with, how often, when, and by what method depends greatly on where I'm going and for how long. If I'm going 60 miles away for a weekend wedding, my communication is much different than if I'm backpacking in India. We all have different relationships with different people in our lives. Like the flowers in a garden, some require more careful tending and more frequent attention. Others will grow and flourish no matter what we do.

If you are traveling on your own and have a spouse or significant other at home, how you communicate is something that you probably won't be able to decide on your own. Your partner will have thoughts and needs about being able to be in touch with you and to know how you are doing. Only you know what works for you and your relationship. Just be sure that whatever plan you make takes into consideration any time changes or difficulties with phone and internet connections and communicate those possible stumbling blocks to your partner.

I am fortunate that I have a very comfortable and supportive relationship with my family and I know that they all like to hear what's happening and how I'm doing when I'm away, but don't necessarily need to be apprised every day as to my whereabouts and well-being. My mother, who happens to be one of my favorite travel partners when I'm not going solo, is different. She has loved and worried about me for over 60 years and I believe is entitled to a bit of extra special assurance when I am away alone.

Our relationships with friends vary greatly and the dynamics of those friendships are therefore different. With some friends, you might be happy to keep them anticipating the outcome and just get together and share stories once you get home. Others you just can't wait to share the day's details with and are excited to email them every night when you get back to your room. It's all up to you.

The secret to this communication thing is, I believe, doing what is comfortable and feels unobtrusive to me. When I travel on my own it is often because of a desire to be alone, to explore not only the world but myself. I travel to escape the everyday pressures of work, bills, responsibilities, and relationships. Therefore, I often don't want to communicate much, if at all, with home. I have to carefully weigh my own needs with understanding the needs and concerns of others in my life. What has helped me is to develop, just like businesses do, a communication plan. Before I leave home, I make a list, either in my head or on paper, of those who just need occasional updates, those who would prefer a little more frequent contact, and those who need regular contact. The secret here, for me, is to decide before I go who is on each list and what I am willing and able to do by way of communication with each group. Some trips I simply tell people that communicating will be difficult, that I'm going to be very busy (they don't need to know with what), and that I will probably communicate only once or twice. Other times I set up group lists so I can write one email, hit send and be done with the whole group. Mom, no matter what, gets her very own communication! (Insert understanding smile here!)

I believe it is also important for us independent and freethinking solo travelers, and that includes you, to remember that the world is a scary place for those who don't travel, those who don't feel the same passion for exploration, and those who love us and don't want to see any harm come to us. They wonder and worry if we're OK. They want to know if we're having fun and feeling good about our travels. Relationships are tricky. They are unique and often complicated, so do yourself and your loved ones a favor and think through this topic with an open mind and heart for all of you.

I can't tell you what you should do, who you should communicate with and how often, but I can tell you that I have found this process to be a whole lot more fun and comfortable if I plan ahead and figure it all out before I leave my front door. For some, depending on the circumstances of relationships, this can be an incredibly difficult part of any journey. We anticipate hearing bad news from home that ruins our trips or we fear listening to yet another lecture about how dangerous a place the world is, or for me, worse yet, I fear hearing about auntie's bunions and grandpa's bursitis and not once being asked how I am doing. I take my traveling peace of mind *very* seriously and do all that I can ahead of time to protect myself and my sanity to the very best of my ability.

Adventure

Remember that list of good reasons and problems when you communicate with home that you created in the Pack Your Bag section earlier in the chapter? Take a look at that list again. Think about whether there is something new or different you want to add or subtract from your list. Once you've done that, work on filling out the information below. Really give this some thought and consider both yourself and your loved ones in making decisions. Let's use your Dream Destination for the purpose of this exercise. Pretend you will be at your Dream Destination for two weeks. Develop a communication plan for yourself that takes care of your needs for being in touch at home. Keep your needs and desires in the forefront. You get to decide how much or little you will be in contact. Fill out the Communication Plan chart below so that you will be prepared.

As usual, I have provided examples to help you get started.

Who to contact:

Person/People in my life	Do I want to/need to communicate while traveling?	How do I want to take care of this communication?
My partner	Yes, every day if possible.	We will use WhatsApp to talk every night.
My coworkers	Only going for 2 weeks. No need to chat.	No communication. Maybe send one postcard to work just for fun.

Looking in the Rearview Mirror

Communication is a tricky travel partner. It changes from day to day, trip to trip, and person to person. Your frame of mind, destination, and relationships greatly influence how you proceed in communicating with home. We all have responsibilities that don't disappear when we leave town. Just like we would with any other travel partner, we need to figure out before our departure what is comfortable, necessary, and what we are willing to do, and communicate that information to those at home.

You will be much happier when traveling if this topic has been addressed and determined before it is needed. Just like many things in our journeys, planning ahead and being able to focus on our travels as we proceed is key to being successful. I'm quite sure that your thoughts about what would make this a perfect trip did not include spending hours on the phone with home or using up your travel days in your room writing postcards.

I'd like you to put this chapter away for a few hours or overnight and think of other things. Once you've had some time away, take another look at the communication plan you created for yourself. See if there are any changes or additional thoughts you've had in the time you were away from it.

Remember, your communication plans will change with every trip, every new relationship, and as your circumstances in life change. It's important to your peace of mind to figure out your communication needs before you set out on your next Journey for One.

Celebration

Well done! You have just passed another checkpoint on your wonderful journey. You have taken control and responsibility for one more element of successful travel.

Now, in celebration of your success and progress, let's put the plan to work. Choose someone who will be a supportive listener and with whom you feel safe enough to share your plan. Get any feedback or ideas that your listener may have about your plan and then adjust the plan as you see fit.

Once you and your listening partner have finished reviewing the plan, celebrate by sharing some time with that person by telling travel stories or sharing travel dreams. Take this opportunity to build your confidence by talking about what you have learned and how your travel dreams are taking shape. Be proud of what you have learned and don't be shy about sharing your success! Congrats to you!

11

EXPLAINING YOUR PLAN TO OTHERS

Why do you go away? So that you can come back. So that you can see the place that you came from with new eyes and different colors. And the people there see you differently, too.

—*A Hat Full of Sky* by Terry Pratchett

Travel Tale

Washington, D.C. was calling, and I just had to go. I hadn't traveled alone very often at that point in time, but was excited to see everything I possibly could in my nation's capital. For my entire life I had seen the sights of D.C. on television, in movies, and in books. I had *always* wanted to experience Washington in the springtime.

My flight was booked, my hotel reserved, and I even knew which full-day tour I was going to join on the first day to get an overview of the entire city. I could not wait to tell my history buff grandfather about my plans. Grandpa's family emigrated from Germany and he was one of the proudest and most patriotic people I'd ever met. I knew that he would be thrilled for me.

Journey for One

I rushed into the house, and after the requisite hugs and kisses I sat Grandma and Grandpa down to share my travel plans. I just knew they would be surprised and that Grandpa would have a million and one suggestions for what I should see and do. Well, I think it's fair to say that I was horribly disappointed. My grandparent's eyes suddenly grew huge, their eyebrows raised, and the smile that had been on their faces slid downward into a heartbreaking frown. I couldn't understand. What was the problem?

"No!" was their immediate response. "It's way too dangerous there for you to be going alone. Don't you listen to the news?" As a matter of fact, I did listen to the news and I was aware of areas of Washington that were better avoided, just like with any big city. I had a good friend who had been raised in Washington and she had said nothing but wonderful things about her hometown and how tourist-friendly it was. She never ever mentioned it being too dangerous to visit. What in the world were my grandparents talking about?

For the next 30 minutes my dear grandparents gave every argument under the sun to get me to cancel my plans. They recalled stories of murders, mayhem, and unbelievable drug problems. They emphasized crime on every corner and that a young woman alone had no business in that city.

I was stunned into silence for several moments, something that is unbelievable to those who know me best. I listened politely as they reeled off their concerns. What could I say? I had no idea how to respond. I wanted to be respectful, but at the same time I was frustrated that they were doing their very best to ruin my excitement. As their arguments slowed to a crawl I realized I was also feeling stupid, like I hadn't done enough research for myself and had paid a lot of money for one of the biggest mistakes of my life. I told my grandparents that I'd have to think about what they said. I would check into their concerns and let them know what I decided.

As soon as I got home I called my friend from Washington and told her about my fiasco with my grandparents. Her immediate reaction was a hearty laugh. She said that my grandparents' reaction was typical. People often get the wrong impression of Washington based solely from news reports on television and in newspapers. We spent quite a while fussing, laughing, and plotting about how to convince my Grandma and Grandpa that I would be safe, happy, and enriched by my trip to D.C. In the end, my friend went with me to visit my grandparents to reassure them. She did quite a good job and even offered for me to have her mother's phone number in case I had an emergency. They seemed to feel much better, apologized for being so negative, and gave me their very hesitant blessing. I would not have canceled this trip just because they had concerns, but I believe it made them feel good to think I would have.

Exploring Washington was wonderful. I thoroughly enjoyed the days in D.C. Were there places I was a little nervous? Of course. Was the city crime-free and completely safe everywhere? Nope. Did I treat it the same as I do anywhere else I travel: with caution, pre-planning, and education? You bet I did. The bottom line is that I was fine.

I learned many lessons about people's reactions when I told them about my solo travel plans. I have realized that most often, their negativity comes from a very well-meaning place of love and concern. I now know that those who are the most concerned are often expressing their own fears and misgivings about how they would feel in that situation. In general, these are the same people who very often have not traveled much themselves. I know that not everyone is going to agree with my plans or approve of my destination, but I do my homework, research the location and its safety, consider my skills, and determine my willingness to perhaps take some risks and to be extra cautious.

Not everyone agrees with my choices about anything in life. Making travel choices and taking adventures on my own is no different. As I continued to travel alone, and my friends and loved ones saw how successful and happy I was, they learned to trust me and my skills. As time went on they had fewer and fewer objections.

Pack Your Bag

Thinking about traveling by yourself is scary on its own. Then, when you think about sharing and explaining your plans with others, it can sometimes become even more frightful. I have shared several times that the key to success for many solo travelers is to plan, plan, plan ahead. With that in mind, take some time to think about a few of your friends and loved ones. How might they react if you were to announce you were traveling alone? Why do you think they might react that way? How can you respond in a way that creates an ally for you where there may have been a critic?

• Journey for One •

Fill out the charts below for two important people in your life. Please choose one who might react in a more negative way and one person who is likely to be supportive and excited for you.

Name of person #1:	How do you think they might react?
Why do you think they will react the way that they do?	How could you respond so that they might be an ally for you?
Name of person #2:	How do you think they might react?
Why do you think they will react the way that they do?	How could you respond so that they might be an ally for you?

11: Explaining Your Plan to Others

Learning the Way

In a perfect world, we would all be adult enough, well-adjusted enough, and confident enough to make a decision about something in our lives and be able to state our decisions to our friends and loved ones without fear of conflict, negativity, or anger. However, we live in a human world where everyone has an opinion, experiences, and fears, all of which they feel a need to share freely.

When I first decided to try traveling alone, I knew myself well enough to know that, while I felt strongly about my decision to travel solo, I did not feel very strong about defending that decision. I was a bit scared and afraid to have it not work out and then have to admit my failure to friends and family. I chose to keep the decision and my practice sessions to myself. This turned out to be good in some ways and not so good in others. You see, when you don't share something about yourself and then you succeed and want to share, you find that you have two conversations that must be woven into one. You first have to give all the background information about your decision and then share the success that you experienced on your travels. That, to me, takes too much mental energy. I am also often an all-or-nothing kinda gal. I thought at one time I would have to share my travel plans with everyone or no one. I went with no one. Like many things in life, it was a matter of trial and error. Once I had been successful a few times and had more confidence in myself I felt much more willing to share and get feedback from those close to me.

As I began to feel more confident in my shoes as a solo traveler I began to explain more often to more people my choice of traveling alone. It was fun to see the amazing number of reactions. A much larger number of people than I had anticipated expressed being envious and wishing they had the courage to do what I was doing. They asked a thousand questions and sometimes, it seemed, were living vicariously through me. Others, as you would expect, would express some form of "Oh, I could *never* do that!" My usual response to that one is, "Oh, OK, then it's good that you don't." In the back of my head, however, I'm thinking that if they really wanted to they would be willing to do the work so they *could* do that. The greatest fun comes when you start to discuss your solo travel plans and discover that the person you are talking to has made solo journeys as well or that they have been to the destination you are planning to visit. It ignites a fire that immediately shows in the smiles on their faces. You have found a kindred spirit. There is so much to learn, so much joy to share, and so much support to be gained by sharing your love of

125

travel. Some of the best information I have gotten about a location, contacts I have made at my destination, or the best travel tips I've learned have come from impromptu discussions like this where I have shared my solo travel plans. I believe that my travels do not begin once I arrive at my destination. My travels begin the minute I start to plan. For me, the planning, anticipation, research, and sharing is half of the joy of a journey. I have learned about myself that sharing my plans only strengthens my resolve to travel confidently and unapologetically on my own.

I would be remiss if I let this conversation end here. We all understand there are people who love us and care about our well-being. When we announce that we are taking off for some new destination, by ourselves, to do who knows what, our loved ones may be concerned. It may be that they think we are too young, too old, not thinking clearly, in danger, not strong enough or any number of other reasons. Perhaps they themselves have never traveled alone so are responding from a place of inner fear of how it would feel to them. Maybe they have a previous experience that leads them to believe this is not a good idea. Perhaps they believe we are being naive or are uninformed about the choice we are making. One option for us it to not tell them what we are doing and where we are going, but that perhaps seems irresponsible and a little dishonest. Could it be possible that they have legitimate concerns about where we are going or what we are doing? We would be immature and childish not to at least consider that possibility for a moment. I have a friend with an incredible desire to travel to North Korea. Depending on what part of the world you are from, this is a dangerous if not impossible goal. Yes, some try to do it illegally, but I wouldn't be a very good friend if I just sat there and said, "Great. You should live your dream, throw aside your common sense, safety, and the legalities. Just go!" I want to warn my friend and tell her what I know without raining on her parade. In those situation, I have found that asking questions about what they know and what they have looked into seems to be taken much better than sounding like a controlling parent. The important thing here is that we need to listen to the concerns of our friends and loved ones. There may be realistic and important information to be gained. It's up to us, after we hear what they have to say, to figure out what to do from that point forward.

There are no magic words of wisdom I can share with you about how to handle this complicated issue of whether or not to share your solo travel dreams. This is one of those choices that you must figure out for yourself. You are the one who gets to choose. You have the power to choose one way and then later choose another. You have the right to do what works for you. How empowering is that?

One of the greatest joys of solo travel for me is that I get to decide everything for myself. I am able to do or not do whatever I want and really spend time being with myself and enjoying life on my terms. If sharing your plans with loved ones makes you happy, do that. If having to explain and defend your choices adds pressure or anxiety to your travel frame of mind, don't do it. It's your decision. You know what will work for you and how to proceed. Just be honest with yourself, and more importantly, trust yourself. It's your journey.

Adventure

"Fake it till you make it" is a line we hear when we are trying to learn to do something that we don't feel very confident or competent about. It asks us to pretend like we know what we are doing, that we are confident and sure of what we are doing, until we come to a place where we really do feel confident and know what we are doing.

For this adventure, I am going to ask you to take a look at the names you wrote down in the Pack Your Bag section of this chapter. Choose one of those people with whom you think you could have an actual successful conversation.

You have three choices for this adventure. You decide which adventure you choose based on what you feel comfortable with at this point and what you think will be the most helpful to you.

Afterward, see how close you came to your prediction of their reactions. How did you do? How did it feel? In what ways were you supported? What, if any, lack of support did you experience and why do you think you you were not supported?

Here are your choices.

- ○ Choose a friend or loved one who you are sure will be supportive of your desire to travel solo. Imagine the reaction you would get from them when you tell them you are thinking of traveling alone. Plan out the conversation and then plan some time with this person and have the real conversation.

OR

- ○ Choose a person you believe may express some concerns or show a lack of support for your desire to travel solo. Imagine the reaction you would get from them when you tell them that you are thinking of traveling alone. Plan out the conversation and then plan some time with this person and have the conversation.

OR

- ○ Choose to talk with both of the people you chose in the Pack Your Bag section. Imagine the reaction you would get from them when you tell them you are thinking of traveling alone. Plan out the conversations and then plan some time with these people and have those conversations.

Looking in the Rearview Mirror

Conversations about something, like traveling alone, that is new for you and perhaps for the person that you are talking with, are difficult. You aren't sure what to expect, you may not be sure of how to begin the conversation, and you may end up being unexpectedly disappointed or supported by someone.

Now that you have had a chance to have a conversation about your traveling alone, how do you feel? Did this activity help you feel stronger about your hopes to travel solo? If you had it to do over again, what would you do differently?

Spend a few minutes reflecting on how this exercise went for you. Then, call someone who supports your travel dreams. Invite them to share this chapter's celebration with you.

Celebration

One of the greatest joys of traveling for many of us is holding the travel brochures or books in your hands and looking at the beautiful color photos thinking, I could be standing there. These brochures make the dream real and give you a great deal of motivation to keep up the work it takes to achieve a successful journey.

Get together with your chosen support person and spend time together collecting information about your Dream Destination. If you are somewhere that is big enough to have travel agencies, go there and ask for information about your destination. These are usually free of charge. Auto clubs, like AAA (American Automobile Association) are also a good source of this information. You are also welcome to go to a bookstore or library to look over the beautiful photos and information about your destination. If none of these options is available where you are, spend time with your friend looking up your destination on the internet. The important thing is that you have someone to talk with while you experience these materials. Enjoy planning, sharing ideas, and dreaming big. No idea is off limits. Just because you share it as an idea doesn't mean you really have to do it. I once came up with the idea of visiting a ranch outside of Las Vegas with ladies for hire just to see what it was like. I never planned on really doing it but it was fun to giggle about. (In the interest of full disclosure... I later actually did go to a ranch like this outside of Las Vegas. It was great fun, but I never went past the gift shop in the front. Yes, I said gift shop!)

Have fun, and kudos to you for exploring this aspect of solo travel. You are getting stronger, more confident, and more skilled every day!

12

MAKING A PLAN

Not planning your travel is "…like leaping off a precipice and trying to knit yourself a parachute on the way down."

—Kelli Jae Baeli (also known as Armchair Detective)

Travel Tale

My good friend, William, had moved to Hiroshima, Japan. I hadn't seen him for a year and I was dying to go for a visit. I had never been to Asia and knew only about four words in Japanese. But I bought the ticket, packed the suitcase and was on my way.

William would be at work when I arrived and would not be able to travel the 30 miles to pick me up at the airport. But we made a plan and I was ready. I was told to get my luggage from baggage claim, go out the door closest to the carousel, find the bus just outside and ask with a Japanese accent, "Hiroshima Basu Centa?" Great! I could do this. I was a full-grown, well-educated woman who had collected several international stamps in her passport. Basu Centa… No problem!

Fast forward to me standing in baggage claim at the Hiroshima Airport alongside about four million teenagers. It turns out that it was Field Trip

Week all over Japan. Every school child in the country was out with their teacher for a week of sightseeing and studying Japan. Luckily, they were all very organized and super well-behaved thanks to their very strict teachers.

I got my suitcase and headed out the nearest door, confidently walked up to the first bus I saw and with my very best Japanese accent asked the driver, "Hiroshima Basu Centa?" with a very question-like rise in my voice at the end of it. "Iie" (no), with an apologetic shake of his head, was the answer. He pointed to the line of buses behind him and I understood that I needed to go farther down the line.

Once again, I got my bags together and headed to the next bus. Nope. Not that one either. Luckily the driver of the third bus in the line-up nodded his head and cheerfully said, "Hai, hai" (yes!). I had found it. I started to climb the steps to enter the bus and the driver then said, "Iie, iie" (no, no), and gestured that he wanted to see my ticket. Ticket? What ticket? He patiently pointed to the bank of machines just inside the door. Yet again, I scooped up my bags and headed inside to purchase my ticket. I stood confidently in front of the machines and then realized that I was in Japan. The machines, the signs, and all of the instructions were in Japanese. *Yikes*! OK, I have choices, I thought to myself. I can go to the airlines ticket counter and change my return ticket to the flight that left for home in two hours or I could put on my big girl panties and do this. All right, Step One. Which machine? As I swung around to grab my bags at the curb I had noticed a big red circle next to the bus door. There was a big red circle on one of the signs overhead so I reasoned that the machine below that sign was probably the right one. Some kind soul had color coded the buttons on the machine so that it was like a stoplight. Green means go, so I went with the thought that this was the one to press first. So far, so good. Step Two. The screen had a list of names, some in Japanese script and some in English script. I was able to make out something that looked close to Hiroshima Basu Centa so I pressed that one. Onward I went. Next step was money. I had no way of knowing how much the ticket was so I used my credit card. I figured it couldn't be a million dollars so I was probably OK. Bing, bang, boom! Out shot a ticket with a red dot on it. Success!

My bags and I marched once more to the bus with the red dot. I showed the man my ticket and he nodded, smiled, and said "Hai, hai." Yes. I was going somewhere. I quickly slid into the seat right behind the driver so that I could ask at the stop if it was my stop. Whew! I took a breath and started to relax a bit. The bus took off and about 10 minutes later pulled into a parking lot with a couple of other buses. It didn't look much like a bus center, but who knows? I leaned forward and again with that questioning lilt at the end of my words, I asked, "Hiroshima Basu Centa?" "Iie, iie," the driver shook his head, indicating it was not, so I leaned back in my seat again. Well, it seems that this bus stops about 100 times on the road between the airport and the city center of Hiroshima. This poor driver after about the third time I asked, finally gestured that he knew where I wanted to go and would tell me when we got to the bus center. I smiled, said my best arigato (thank you), and waited ... and waited ... and waited.

After about an hour we pulled into a location that clearly was a major bus center and I saw the name Hiroshima over the main entrance. I was sure this was it. Just about that time the driver turned and started to say something. He was joined by every single one of the 40 other people on the bus in a hearty chorus of "Hiroshima Basu Centa!" They were all going to make sure I knew I had made it. I bowed, like I'd seen in Japanese movies, said my arigatos, grabbed my bags and entered the main station thrilled to see my friend standing there waiting for me.

Hiroshima was amazing and I loved Japan; the food, the people, and the Hiroshima Basu Centa! The plan had worked. I knew what I was going to do once I got to the airport. I'm pretty laid back and take things as they come, but when I'm alone and doing something like this on my own I have found that — just as important as my toothbrush, guidebook, and camera — is my plan. I make sure I have one. Just in case I need it, I've got it. It gives me confidence and comfort to know what, when, and how my travels should play out.

Pack Your Bag

What things in life do you make a plan for? Myself, I'm a list maker, which to me is a plan of sorts. As a teacher, I made daily lesson plans for years. I had to write out what I wanted to accomplish, how I was going to do it, what I needed, and how I would know if I'd been successful. Most things in our day-to-day life don't require that much detail but traveling alone is definitely one of those things, especially when it is new for you, that is much more successful when it is well planned.

Some people are planners naturally, others are not. No problem. When you are traveling on your own you will find a method of traveling that works for you. If you are not one of those natural planners, I am going to ask you to stretch yourself a bit here and go with me. Once you've done it and made your way through this book, then you are more than welcome to plan your travels, or not, any way you choose.

Just like with my lessons needing plans for the kids, I believe that when you travel it is important to have a plan.

Learning the Way

The word "plan" gets tossed around a great deal in today's world. We, as a society, can plan everything from the birthdate and gender of our babies to pre-planning our funerals. We have diet plans, health plans, construction plans, and business plans. Are we all that busy that we need a plan to keep our lives in order? Well, for some of us, including me, the answer is yes. But when I travel and make a plan, it's not with the purpose of remembering as much as it is a tool for organizing and directing my time and resources. Whether it is a business plan or a diet plan, the thought behind it is to clearly lay out what you should be doing, what you want to accomplish, and making choices about your time and money to get the best results. Results require planning. Planning requires goals and goals require a target or expected outcome.

You would never get on an airplane without knowing for sure where it was going. You wouldn't live on celery and tuna fish without knowing what you might be able to expect in terms of possible weight loss. Similarly, you would never set out on a vacation without knowing where you wanted to go, what you wanted to do, see, or experience. It's also important to find out how to do things, how much it costs, and what you can expect in terms of what you want to do, see, or experience. Take a look at the following example of planning for a trip to Chicago.

This is when I really start to get excited. I get to plan! It's the same feeling I used to get when the Sears Christmas Catalog would come out when I was a kid. There were pages and pages of gift potential between those covers. Suddenly, I turn into the little 8-year-old girl who spends hours penning her letter to Santa. I can ask for *anything* I want. After that it is Santa's job to figure out if there is enough time and money to make it happen but until then, the sky's the limit. You get to fill those days and plan for those wonderful destinations.

When am I going? May 3-10	Where am I going? Chicago, Illinois, USA	
Name of what I want to see.	What is important to me to see, do, experience there?	How much time do I need there?
Lake Michigan Tour	Architecture Old Navy Training Center Overview of downtown buildings	15 minutes tour 10-20 min. wait 1 hour total
Skydeck	Top of 103 story building View of 50 miles and 4 states Ride glass elevator	Hours 9am-10pm 2 hours on top 1-2 hours in line 4 hours total

Aside from planning for the sites you want to see, don't forget about planning for travel time. It takes time to drive to the airport, take a plane, get to the hotel, find your way to sights, eat meals, and wait in lines. Those things, along with not checking the opening/closing hours of a site, are the biggest troublemakers in vacation planning.

To avoid the occasional scheduling mishap it is good to get information about your travel arrangements. Thanks to the internet we can now find out just about anything we could ever want to know. "Alexa, how long does it take to drive from the airport to the Happy Traveler Hotel during rush hour?" Find out layover times. Relax and expect the unexpected long wait at the rental car agency or passport line. You'll be fine if you've already added a bit of time into your schedule for such things.

I also encourage you to build in some free time, time to relax, time to enjoy an unexpected attraction, shop, or to spend with a new acquaintance you meet while traveling. It's also a great time to make up for things you had to miss because of over-planned days, weather, or waiting in lines.

How you create your plan is up to you. While some travelers enjoy a more freewheeling and open-ended way of traveling, for beginning solo travelers I have found that this leads to difficulty. Many of the fears first-time solo travelers have seem to come to life during unplanned and unstructured time. If you have a plan full of things to do, places to go, and experiences to look forward to, you are much less likely to feel loneliness, homesickness, and the sense of not knowing what to do, than if you had a plan.

When choosing the type of plan you want and need, think about what plans you have used in the past, your familiarity with the place you are going, and the amount of changes in activity you will have during your day. I typically have a much more detailed plan on travel days because of having flights that are scheduled at certain times, hotels that have specific check-in times, and public transportation like buses, trains, and subways that won't wait for me if I'm late. It's important to keep in mind, however, that while they won't wait, you can usually get a ticket for a later departure. But why would you want to go to that hassle if all you need to do is plan in a more detailed fashion and avoid the problem? Sightseeing days and relaxation days do not need to be as carefully planned unless you are seeing many different sites or are simply the kind of person who feels more comfortable knowing exactly what is planned at every moment.

Let's take a look at three different types of plans: the General Plan, the Detailed Plan, and the In-Depth Plan.

General Plan

The General Plan is exactly as the name implies, general. It is a brief overview of the highlights of your day. It might be arranged by preference, by opening times or suggested order, or it may be arranged by transportation needs such as a hop-on hop-off tour bus. It's a simple way to make sure that you don't miss something you really want to do. Here is an example of a General Plan.

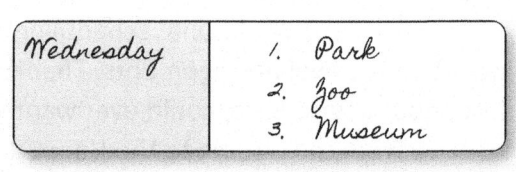

Wednesday	1. Park
	2. Zoo
	3. Museum

Detailed Plan

The Detailed Plan contains more specific information. It generally contains times and details that may be necessary to make the rest of the plan work.

• 12: Making a Plan •

For example, the plan below lists getting a bagel and coffee on the way to the City Park. This may be a detail that is important so you can get to the park on time for an activity, or it could be listed because the idea of getting a bagel and coffee from a food cart near the park is an experience you want to do for fun. Another example on this plan is going to lunch at a specific restaurant. You may want to list it this way because you are meeting someone there and don't want to be late for your meeting time, or it could be that this restaurant has been suggested to you by a friend and you want to be sure you have the right name and address at hand while you are touring that day. Another detail that can be helpful in this type of plan is a note at the end about what is happening the next day so that you can prepare for what time you want to go to bed, what time to get up the next morning, or figuring out how late to stay out that night. I find this type of plan particularly helpful on travel days. This helps me keep times organized such as flights, shuttles, meetings, tours, or activities that occur only at a given time.

> **Thursday**
>
> 9:00 Go to City Park, walking.
> Get bagel and coffee on the way.
> Go to see the duck feeding at 10:30.
> Catch crosstown bus to restaurant.
>
> Have lunch at Tony Bloom's restaurant on East 69th.
>
> 1:00 Get to Jamboree Theater by 1:30 to see a play.
>
> 4:00 Take subway back to the hotel to clean up and rest.
>
> 7:00 Dinner with Ellyk family.
>
> 10:00 Back to hotel to sleep. Early morning airport shuttle.

In-Depth Plan
Traveling that includes a great number of activities or has several specifically timed activities I usually schedule with an In-Depth plan. This type of planning

prevents missing start times or having to backtrack because of poor planning when I travel. If I am planning to be in a location for only one day or a very short time, I will use the In-Depth planner so that I make the best use of my time and have less chance of forgetting or missing something I really want to do.

In-Depth Plan

Thursday	8:00 Alarm
	9:00 Breakfast in hotel restaurant
	9:45 Catch bus 48C on the corner of 70th and Main St.
	10:15 Arrive at Museum of Textile Arts, 416 Nederly Ave.
	10:30 Start Folkloric Weavers Tour
	11:30 End of Tour
	12:00 Have lunch at Shelneri Deli, north side of museum
	1:30 Back to the hotel by cab for rest and relaxation
	4:00 Get ready for art show. Dress is dressy casual
	5:00 Get taxi at front of hotel, go to M & E Pottery Gallery 1227 De Braun Circle
	6:00 Opening words and show begins
	7:00 Spend time looking at art and mingling with others
	9:00 Taxi to Cafe Augustus for dinner and classical concert, 47 Salamander Street
	???? Horse carriage ride to hotel with Golden Carts Co.

Which type of plan you use is far less important than the fact that you make a plan. You have invested a great deal of time, money, and emotion in this trip. Why would you want to take a chance on missing something important or spending all day on a bus just trying to find a museum you want to see? Your time will be much better spent before you ever leave home by researching sights to see, schedules, options, and transportation. Don't waste valuable time being lost, confused, or frustrated. Especially on your first solo trip, plan well so that you can greatly reduce your anxiety and instead spend your energy and your time having fun. Your first solo trip is more about having the experience, trying out your wings, and building your confidence than it is about seeing everything.

Only you know your interests, your passions, your preferences, and your budget. You get to plan a trip that is exactly what you want it to be. If you under-schedule, you run the risk of having too much time on your hands, which leads to boredom or loneliness. If you overschedule yourself, you may become exhausted, frustrated, and feel defeated, resulting in a negative first solo journey. This is not what anybody wants for a vacation. So, get out your computer, open up your notebook and start planning. You are the master of your travel destiny!

Adventure

Now that you know the planning options available to you, plan a half-day excursion for yourself. Choose two to three things that you've never done before. Your plan can be very simple or more complex, depending on the amount and types of activities you have chosen, your personal style, and your level of comfort with going out and about on your own. I'm giving you a space to do your planning here but if you have another format or method that works for you, use it. You are the travel guide on this trip. You're the one in charge. Get going. You're going to be great!

When am I going?	
Where am I going?	
Here's my plan for the things I want to see, do, and experience...	

Looking in the Rearview Mirror

There are several ways I evaluate the success of my travels. One of my favorites is my **1, 2, 3, A, B, C Review**. Here's how it works:

What is the **one** thing that was the absolute best part of my journey?
What **two** things would I do differently next time?
What **three** places, activities, and/or experiences do I want to tell others about?
A = Anxiety: What was my anxiety level and what did I do to calm myself?
B = Bravery: What was something that I encountered that made me depend on my bravery?
C = Confidence: What is something that caused me to have more confidence in myself?

Celebration

Before you start this celebration I have a little gift for you. Going to a restaurant and eating alone is often one of the most difficult things that a first-time solo traveler must do. Part of the reason it can be so unsettling is because we don't know what to do to keep ourselves comfortable. My gift to you is a list of things that you can do to occupy your mind and your time while you are dining alone. You will see how much easier it is to enjoy a solo restaurant experience when you take some ideas and tools with you when you go. These ideas will make you feel more comfortable and help fill the void of not having someone to chat with while you eat.

Choose a restaurant in your city or area. It is best if you choose one that is fairly busy. This often makes first-timers feel a little more comfortable and less conspicuous. I encourage you to go somewhere that you have not been before so that it is more like a solo travel experience. Now, look at the goal plan below and prepare for your truly enjoyable meal out.

Don't forget to jot down a few notes about how the experience went for you when you get home. Do it as soon as you can. Time has a way of dimming our memories a bit. Bon appétit!

Solo Restaurant Adventure Goals

What do I want to see, do, or experience?	What are my concerns, fears, and hopes?
Eat alone in a restaurant.	

♦ Journey for One ♦

What can I do to make it more comfortable for myself?	How did this experience turn out?
○ Take a something along to read like a book or magazine. ○ Take along something to write. ○ People watch, notice how many other people are eating alone. ○ Do something on my phone. ○ Dress comfortably so I can feel more relaxed. ○ Play car games in my head, like ABC games, first letter = last letter game; counting things (like the number of people with glasses), anything to keep my mind occupied and entertained. ○ Take along earphones and listen to music, audiobook, or podcasts. ○ Make up stories about my fellow diners and the details of their lives that bring them to the restaurant. ○ Pay attention to all of the things that I like or enjoy rather than the things that might make me uneasy.	

13

BEING FLEXIBLE

Blessed are the flexible, for they will not be bent out of shape.

—Author Unknown

Travel Tale

My current bucket list has about nine thousand things on it. OK, maybe not that many, but certainly quite a few places, cultures, or events that I truly want to experience. One wish that had been on the top of my list for a very long time was a hot air balloon ride. After moving to Austria, I began dreaming of a ride over the Alps as my preferred location for this trip.

I refuse to spend my life hoping, wishing, or thinking "maybe someday." If there is something I want to do, I make it happen. I have no intention of leaving this life with a lot of "I wish I hads." I wish I had gone to Antarctica or I wish I had eaten dumplings in Korea. If I want it I find a way to make my wish come true. No fairy godmothers for me.

In keeping with this tradition, I soon chose my favorite lake location in the Alps and began researching balloon companies in that area. I finally found what I was looking for; it would be perfect. I called the owner of the company,

◆ Journey for One ◆

gave her my deposit, and scheduled the ride. I had two months to wait, but it was going to happen.

As the date for my ride came up I got more and more excited. I bought a new camera so I would have amazing photos from my ride. I wrote and rewrote my packing list at least a hundred times. Finally, the time to sail the skies over Lake Attersee was here. I packed my bag and set out the day before to the gorgeous surroundings of the lake and spent the night before my ride in a charming, traditional old guesthouse on the shore of the lake. I called the balloon's pilot, Ingrid, and left a voice message that I was in town and ready for my flight the next morning. I asked her to call me back and give me details of where to meet and at what time. I settled into a restful afternoon of swimming in the lake and sipping wine at the water's edge.

After a few hours, I began to be a bit frustrated because Ingrid hadn't called me back yet. I knew this was just another day in the air to her, but to me it was a big deal. I called again and left another message, and another, and another. Late into the night I called. Never a response. I barely slept all night, switching from being incredibly angry to intensely disappointed. I had gone to so much trouble to make this a perfect flight and now it looked like it most likely wasn't going to happen. By 9 in the morning I knew that there was no hope. Balloon flights there had to prepare before dawn to catch just the right winds and start liftoff at sunrise. I was devastated.

I normally don't anger easily, but in this case, I was fuming. I started off down the road, headed for home, and finally realized that I could either pout or I could find a nice way to spend the two days I had left on the trip. What I chose was to spend the two days in an area of Austria called the Salzkammergut. It's a beautiful mountain area in the Alps that is famous for its breathtaking scenery, picturesque traditional villages, and the numerous salt mines that are scattered around the area. I ended up spending two days in absolute bliss. It was exactly what my disappointed little heart needed. Not only did I prevent myself from going home and having a pity party, but I found a place that became my favorite area in Austria. It is where I return when I need to breathe, to escape, and to find my joy. Had the problem with the balloon ride not happened, I would have never found my happy place.

Two weeks passed before I ever heard from Ingrid. When she finally called, she started the conversation by apologizing profusely. She told me that she had been in a horrible car accident on the way back to Attersee from a balloon festival and had been taken by ambulance to the hospital with serious back injuries. Her computer, which had all of my contact information on it, had been thrown out of the car in the accident, and for two weeks had sat in the ditch after the car was towed away. After getting her wits about her again after the accident, Ingrid sent her husband and kids out to the accident site to find her computer. She said she was beside herself because she knew she had scheduled a balloon ride with me but had no way to get in touch to tell me what had happened. Her family finally found the computer and then my contact information and that's when I finally got the call. My anger disappeared completely and I became concerned for the pilot and was relieved to hear she was going to be all right.

Before ending our conversation, Ingrid told me that she wanted to make my dream trip happen and that to make up for disappointing me she was going to give me the balloon ride at no cost. She insisted that I return to Attersee and ride with her into the sunrise on the morning of my choosing. I have not set the date yet for my balloon journey, but you can bet that I will enjoy every minute of it.

Pack Your Bag

It's time for a little self-assessment. Think about yourself and your style of travel at this point in your life. Are you the type of person who has fairly solid hopes and needs when you travel or are you comfortable with the little twists and turns that traveling can present? Once you've thought about your own personality in terms of change, I'd like you to spend some time writing out a story of a time when you had something go wrong during travel and how you dealt with the situation. Don't forget to include information about how it turned out and your overall feelings about that trip. Have fun remembering!

_____

_____

_____

_____

Learning the Way

Are you old enough to remember a candy bar called a Big Hunk? It was popular in the U.S. during the 1950s, '60s and '70s. It was a big, long strip of honey sweetened nougat with peanuts in it. One of the things I liked best about eating a Big Hunk was that it was chewy and flexible and you could stretch it, twist it, and pull it until you got just the perfect bite of chewy yumminess. It was fabulous. My choices in travel are very similar to my taste in candy. I truly enjoy experiences that stretch me, that make me turn and twist a bit to get just the morsel of life I want. I can decide how big or how tough I want each bite to be, but I know in the end that the result will be well worth the effort. The only way to have travel experiences like that is to be flexible, be willing to go with the flow, and to throw out the phrase "but I'm not used to that."

One of the main purposes of travel is to gain new experience. When I travel I want to see new things, meet new people, and see life from a different perspective. If I do only what is comfortable and what I am "used to," then why leave home? Don't misunderstand me, I want to be comfortable on vacation, but not in the way that we talk about experiencing your destination. I want to arrive home after traveling changed, different, more. It's like the old saying, "If you always do what you've always done, then you'll always get

♦ 13: Being Flexible ♦

what you've always gotten." Besides my interest in being adventurous about my traveling life, I feel that I am being incredibly insensitive, even rude, if I travel to a new place and expect for everything to be the way I like it at home. Would I be happy if guests from another country came to my city, my home, and judged it based on how much it is not like their home? Certainly not. Therefore, I do my very best to respect the differences of a new destination rather than expecting a photocopy of my home life.

Being flexible is the key to receiving the gift of great experiences. The majority of wonderful memories I have of my travels are things that were unexpected, unplanned, and unknown to me when I left home. We all make our plans for our travels and hope with all our hearts that things go according to that plan. But, if you really listen to someone else describe their vacation, they never say, "It was great, the plane was on time, the hotel bed was perfect, I loved every morsel food that I ate, and every site that I saw exceeded my expectations." No! The joy in our travels and our very best stories come from those unexpected, completely off-the-plan moments that inevitably happen every day. We find joy and even humor in some of the most silly and ridiculous circumstances that we encounter on our travels. Let's try to look at those "happenstances" as gifts and opportunities to stretch rather than as annoyances or struggles. Let me give you an example.

One of my most enriching cultural experiences came from one of those, "Oops, this isn't what I had planned" moments. I was visiting Lisbon, Portugal for a week. One of my favorite things to do when I travel is to find and enjoy the local musical culture. Portugal has a wonderful musical genre called Fado. It's kind of like Spanish Flamenco music and dance. I badly wanted to see it live while I was in Lisbon. So, I did my research, found what I thought would be the best show, best price, and best location and ordered my ticket for later that night. I got in a taxi, went to the club, and experienced a lovely dinner show. I was a happy girl. At the end of the evening I went out on the street to hail a taxi and found that the streets were pretty empty and there didn't seem to be any taxis anywhere. I decided I would stop in the next restaurant or bar and ask if they would be kind enough to call for a taxi for me. Understand, I don't speak a word of Portuguese, but as I've discovered, there is *always* someone who speaks English or is willing to try. The gentleman who was helping me was dressed in traditional Fado attire and we started to chat. He asked if I had enjoyed my evening and if I liked Fado. With great enthusiasm,

◆ Journey for One ◆

I told him how much I loved the music and how I could feel it in my bones. He was so pleased that I liked it that he asked me if I'd like to stay. This was a local Fado club and they were just getting ready to start their next show. I was thrilled. Are you kidding? Of course, I wanted to see their show! My new friend, Fausto, showed me to a table of his friends, and so began one of the best experiences of my life. You see, the group of friends I sat with were all Fado performers and as we sat and drank way too many glasses of port and listened to the intense emotion of the Fado, I got stories and information, and even instruction on Fado. It was magical, a night I will never forget, all because I was flexible. Rather than being upset because I couldn't find a taxi, insisting on going directly to my hotel, or refusing Fausto's kind invitation, I was flexible and a whole new experience opened up in front of me. I am so grateful I am that I went with the flow and allowed flexibility to be my guide.

One of the most paralyzing phrases a traveler can use is "I'm just not used to ... " Depending on how you say it and the intention behind it, saying you're not used to something is like giving yourself permission to not try. It is often followed by the unspoken thought that "my way is the best or only way."

Instead of focusing on the negative by saying or thinking "I'm *not*..." try turning it around to "I'm flexible. I'm open to new options." You'll find that just by opening up your mind to new possibilities and experiences that your travels and day-to-day experiences while traveling suddenly take on a richer array of options. I'm assuming that because you picked up this book you are interested in diverse perspectives, new adventures, and in broadening your own horizons. Just keeping yourself open to the idea that the unexpected is an opportunity rather than being an obstruction gets you ready to travel in a new and more confident way.

When you have expectations that something will be a certain way and you set your whole sense of whether you've had a successful day or trip on having everything turn out exactly as you planned, it closes the door to something even better. I remember going to a nice restaurant with friends and seeing that one of their specialties was oysters on the half shell. I love oysters. I love them fried, boiled, smoked, but raw? I don't think so. It's always been a texture thing. Well, shaming myself a bit for saying I didn't want to try them was just like saying "I'm not used to ... " or, "I'm not going to ... " when I travel. Needless to say, I convinced myself that it wouldn't hurt to try. If I liked them, great. If I didn't, it wasn't going to kill me. My friend and I ordered six. I'll just say that my friend enjoyed five raw oysters that night. I found out through this little experience that I was not, indeed, a fan of fresh, raw oysters, but being able to have this friendly, funny little experience with my friend was priceless. This is the gift of flexibility.

I'm not saying that you shouldn't have expectations or preferences. You don't have to do, or like, or try everything. I'm just suggesting that when you are presented with something that is different and you have the opportunity to try something new, stop and give it a few moments of thought. Consider what benefit there might be in giving it a try. Weigh the pros and cons. Think about what you could gain by giving up your plan to do something different.

There are times during travel that we run into complications that truly cause great difficulty. We deal with those and do the best we can. Sometimes flights get delayed or canceled. Perhaps we have to stay overnight in an unplanned city. This can be troublesome because of family, other commitments, connections, or cost. Once in a while we have no other choice than to go with the flow and accept the conditions as they are. We still have options about how we react to the situation. Accepting the situation as it is can be an opportunity for flexibility. I am pretty good at living with changes in my life, but one of the things that makes me cringe is when, because of circumstances beyond my control, my travel is interrupted by a canceled flight, requiring me to spend a night in an unplanned city. It is usually a result of bad weather conditions and thankfully doesn't happen often, but it has happened three or four times that I can remember over the past 47 years. I think it's frustrating because I've got my head all set for either leaving or arriving home. I want to get where I'm going hassle free.

The first time my travels were interrupted I was frustrated, angry, and in hindsight was more like a spoiled brat than a well-traveled adult. I sat in my room and sulked, went to the hotel restaurant for dinner, and then sat in my room plotting what to say in the scathing letter I was going to write to the airlines. What energy that took! I didn't sleep, I felt terrible, and it almost eliminated memories of an otherwise delightful journey. Several years later when it happened again, I thought, OK, I'm stuck in this city, far away from the city center, and there's no shopping or restaurants within 10 miles. I can either sit and plot revenge against the airlines, which is not the version of myself I want to hang out with, or I can use this opportunity as a spa retreat. I chose the latter. I called room service and ordered a glass of wine, ran a nice hot bath, got a magazine from the hotel lobby, turned on some peaceful music on my phone and, voila, instant spa. Afterwards I crawled into a bed I didn't have to make, propped myself up with the four puffy pillows, and turned on a movie of my choosing. I know a lot of women who would give anything to have an evening like that. See what I mean about having flexibility in your thinking? It can totally change a travel experience from being a disaster to being an extraordinary opportunity.

So, what I am saying is that it's like my favorite grandpa used to say, "If you want everything to be like home, stay home!" If you are ready for new and exciting possibilities, put on your flexibility suit and get traveling.

Adventure

It's time to get out and about on your own again. One of the experiences that people find very uncomfortable when being alone is going to a sporting event, movie, concert, or other performance alone. I can hear you now. Noooo ... you are yelling. "I'm not used ...", oops, I mean, "I'm flexible and I'm open to new options!" Find an event that interests you and make plans to go there on your own. Keep in mind the things that you have learned already and set out confidently. I realize it may not be easy, but here are a few tips from me:

- Start out by paying attention to how many people are actually alone besides you. It doesn't take long to realize that you are not the only one attending by yourself.
- Don't arrive too early. The later you arrive, the less time you have to sit there alone. It is only the brief time before the movie starts that is uncomfortable. Once the lights go out, nobody is paying attention to anything but the performance.
- It helps when you feel like you are being stared at to try to remember that there are a million and one things the person staring could be thinking about. This means that there's a 999,999 to one chance that they are not really thinking about you.
- If the event you are attending has a program, spend lots of time reading it, even if you're really not *that* interested. It keeps you busy and less anxious.

○ Don't get so stressed out about being alone that you forget to enjoy the performance.
○ Have fun!

Looking in the Rearview Mirror

You did it! You made it through the adventure of going to a performance on your own. Congratulations! How did it go? Spend some time reflecting on your thoughts while you were sitting there. Did you focus on the performance or were you completely occupied with being alone? Answer the questions below for yourself.

On a scale of 1 to 10, 10 being the highest, by the end of the performance how comfortable were you with being alone? _____

What things were pleasant about being on your own?

_____
_____
_____
_____
_____
_____
_____
_____

What things made you feel uncomfortable at first?

_____
_____

♦ Journey for One ♦

_____
_____
_____
_____
_____
_____
_____

Were you feeling better by the end? If yes, in what ways? If no, what still felt uncomfortable?

_____
_____
_____
_____
_____
_____
_____
_____
_____
_____
_____

♦ 13: Being Flexible ♦

What would you differently or do again to helpy you feel more comfortable the next time you are at a performance on your own?

_____
_____
_____
_____
_____
_____
_____
_____
_____
_____

Celebration

One of the things that travelers have in common whether they travel alone or with others is that they love to share their travel stories. Often times all you have to do is say, "Hey, tell me about your last vacation," and they are off and running. What you are going to do now is find a friend or family member who travels, sit down with them for a cup of tea, coffee, a glass of wine, or a meal and ask them to tell you some of their stories about times when things didn't quite go as expected. Get them to tell you what happened, what they did about it, what they did instead, and how it ended up. Pay attention to how they tell the story. Do they tell it as a horror story or do they relate it with a sense of humor and as almost a fond memory? Obviously, this depends on the person and their personality, but I find that it's the mishaps and unexpected that become the greatest travel stories. See if this isn't true for your friends and fellow travelers as well. Thank your friend for sharing their experiences and helping you build your courage and skills.

14

FEAR

It's believing in our dreams and facing our fears head on that allows us to live our lives beyond limits.

—Amy Purdy

Travel Tale

I remember lying in bed the night before a week-long rafting trip down the Colorado River. I was scared out of my mind. I had friends who were experienced and certified river guides leading the trip and I trusted them. What I didn't trust was the river. Why in the world had I ever agreed to make this ridiculous trip?

This particular stretch of the Colorado River we were going to navigate required a special license to raft it. It is considered a Class 5 river, deemed extremely dangerous. The most difficult and dangerous rivers are considered Class 6. In order to get the required license, we had to have a certified rafting guide in charge of our raft. We had met all of the requirements but somehow that didn't make me feel any better.

I just knew I was going to die. I was young, healthy, and had a whole wonderful life ahead of me. I cried, shook, and even threw up as I laid awake the entire night. What was I thinking? This was stupid. I ran through every possible scenario of what could happen to me. I could be thrown out of the boat, hit my head on a rock and die. I could be attacked by a bear as I slept on the bank of the river at night. The walls of the canyon could suddenly give way and I could be killed by a giant boulder. Never, in all of my horrendous scenarios, did I end up enjoying myself and being grateful that I had come. "It will be fine," they said. "It's all good. We've done it a hundred times before and been fine." Why then was I filled with such dread? Why didn't I just listen to my gut and stay home?

Well, as I later discovered, my gut lies. My gut is a selfish little creature. It is not a rational thing. It seems to take whatever it sees on television, hears on the news, or dreams up in its stupid little mind and insists on it being the truth. Sometimes. This river trip though, I just *knew* was for sure destined for doom.

I can't remember why I didn't call and cancel at the last minute but I'm sure it had something to do with not wanting to disappoint my friends, and I had paid the full amount for the trip so I couldn't let that go to waste. I figured I'd just have to do it. I'd just have to float that river and die. I said heartfelt goodbyes to my loved ones, gave an extra-long hug to the dog and set out for our rendezvous point. I cursed at myself the whole way there for not having the guts to back out.

I arrived at the designated parking lot, unpacked my gear and began to relax a little as I chatted with my friends and saw how excited, relaxed, and happy they were. Surely, they didn't want to die and wouldn't have done this numerous times if it was that bad. Still scared out of my mind, I loaded my gear onto the raft and we set off down the river watching the hawks, which I was sure were actually vultures, soaring overhead.

First hour: OK, I'm still alive. Second hour: OK, maybe I won't die today. Then I realized it had been half a day and I was actually smiling. I was beginning to think I could have fun, at least until I died. Then it happened — our first set of rapids. This was it. I just knew it. I grabbed every piece of rope, rubber, equipment and hope that I could and prepared to meet my maker. About 30 seconds later we started over the rapids, rolling to and fro as we dodged rocks and valleys in the rushing water. Suddenly we were gliding gently down the river again. I opened my eyes and realized I wasn't dead. I had survived. After I started breathing again, I caught myself laughing. Hey,

◆ 14: Fear ◆

that was kinda fun, I thought. "Are there more like that?" I asked my friend with the oar. "Oh yeah, we get into some bigger ones later this afternoon." Then, from out of nowhere, I heard my voice say out loud, "Cool!" Cool? Twenty-four hours ago, I was crying and throwing up and now I think these killer rapids are cool? Indeed, I thought they were cool and they became more and more cool as the days went on. The second day I found myself sitting at the front of the raft, holding onto nothing more than a little rope, and bending forward and backward trying to get a little more motion as we navigated another of the river's numerous rapids on that trip.

That first night, as I lay in my sleeping bag counting the hundreds of falling stars in the sky, I thought to myself, Why was I so scared? Why was I so certain that this was going to be bad and that I would die? At the time, I had no idea why I'd had such dismal thoughts, but I knew that I had wasted a whole lot of time, emotion, and energy on something that never happened. Years later I read a quote from Les Brown, "Too many of us are not living our dreams because we are living our fears." That was it. Why was I giving fear more power in my mind than hope or desire or joy? Fear, I decided, was somewhere I did not want to live. Living in my hopes and dreams, served with a small side of caution made from fear, was a much better choice.

I woke up the next morning in a newer, brighter, and more adventurous state of mind. I couldn't wait to get back on the river and see what the day had to bring. I was now looking forward to the rapids, the floating, and whatever surprises the day might bring. Two days later we hit the big rapids of Cataract Canyon and we all screamed and laughed and hooped and hollered as we bounced up, down, and sideways in our little rubber rafts. I couldn't ever remember having so much fun and feeling so alive. I can honestly say I was truly sad when the trip came to an end. I couldn't wait to get home and plan another rafting adventure.

Undoubtedly, my most treasured souvenir from that trip was the knowledge that yes, I get afraid sometimes, but it's just like a cold. I feel yucky, I want to stay in bed, but I know it's temporary and I just have to take care of how I feel and push through until I'm on the other side of the cold and feeling great again. I don't have to live with that cold forever.

I have rafted down many rivers since that first trip, all amazing, but I think the journey that took me the farthest in understanding myself and finding joy in traveling on my own was that wonderful float down the Colorado River.

Pack Your Bag

We are all afraid of something. Sometimes that fear is a little uneasiness, like when you are afraid that you won't have enough candy to pass out on Halloween. Other times it is a great big fear, like when you're afraid you are going to die on a rafting trip. A huge fear for one person may be nothing at all to another. None of us can judge another's fear. We can't say "Your fear isn't as bad as mine," or "You're afraid of that little thing? How silly!" The fear we feel inside is different for each one of us. Our experiences, personalities, upbringing, education, religion, and life choices all contribute to whatever it is that causes fear for us.

For this exercise, I want you to put away all judgment, all self-criticism, and instead open your heart and your mind so that you are able to be very honest with yourself. Nobody is watching. No one is going to judge your answers. Just write down what you actually feel.

Think about a journey to your Dream Destination. You've written about what you would see and do there. You've dreamed of how you would get there, where you would stay, and what would make you happy on that trip. Now I want you to write out a list of things that are frightening to you. What parts of that journey are you scared of? Again, there are no wrong answers, just complete honesty here. Feel free to elaborate as much as you like so that you remember very well what it is that frightens you. I have done one in the first box for you as an example. Don't worry about the Level of Fear Column for now. Just leave it blank and you will fill it in later.

Dream Destination:

Level of Fear	I am fearful about …	Explanation, clarification, and exploration of where this fear might be coming from …
	Being around cats. I hear they have lots of cats there.	I was bitten by a cat when I was 6 and am very afraid of cats, especially wild ones.
	There will be a lot of stairs to climb.	I have bad knees that only hurt when I have to go up or down stairs.

Level of Fear	I am fearful about ...	Explanation, clarification, and exploration of where this fear might be coming from ...

As you engaged in writing your list of fears above, I am fairly certain that certain fears seemed a bit more out of your comfort zone than others. That's perfectly normal and we all have those same feelings, but you did it! You identified your fears, clarified them for yourself, and now you know what things you have to face in order to be ready for that Dream Destination. Now it's time to go on to the next step.

Not all fears are created equal. Some fears, true fears, scare the socks off of us to the point that we are almost paralyzed by the fear. Other fears are just little things we're not familiar with or not sure how to deal with. Those things I would call worries. Below is a fear continuum. It gives a number to the different levels of fear between something that is a just a worry, all the way to a true fear.

1	2	3
A Worry	**A Worrisome Fear**	**A True Fear**
These fears are something I try to avoid.	These fears scare me but if I knew how to deal with them I'd be OK.	These fears scare me to the point of wanting to run away.

Once you've looked at this continuum I'd like you to go back up to the Pack Your Bag section and label each of your fears as either a 1, 2, or 3 depending on how you feel about that particular fear. Come back to the Learning the Way section once you have done that.

Learning the Way

Every day each of us makes a million decisions. We decide what to wear, which way to go to work, what to make for dinner, and whether we want to go to a movie or play putt-putt golf. Some of those decisions come more easily than others. Some require deeper thinking, more knowledge, or more consideration than others. For example, most days deciding what to wear to work is a no-brainer. You make that decision fairly easily. But on the day that the company president is coming in to do an evaluation of your department, you will probably struggle to figure out what to wear because some amount of fear comes into the picture, and making the right decision is much more important.

The same is true for travel. You might be able to drive two hours down the highway on your own to visit a friend and not think a thing about it. But deciding to get on a plane and fly to a jungle island for 10 days is, as they say, a whole different ball game. Fear can take over your thinking. The decision now contains a significant element of fear, the outcome is much less predictable, and the planning and preparation are much more important.

When we decide to travel, whether alone or not, fear becomes a much bigger part of the equation. It brings up thoughts of safety, danger, communication, isolation, and unfamiliarity. Naturally, the level of fear is much greater than with our day-to-day decisions. We have become so used to making decisions on just a brief thought or no thought at all that we forget that not all decisions are created equal. Sometimes a decision requires more information, more research, and more insight than what we might possess on the spot.

I recently refinanced my house. I am not well-versed on financial topics, I don't enjoy dealing with money matters, and I have what I jokingly call finance phobia. I rely greatly on professionals to help me manage my money. I acknowledge my challenge, face the fear, and research a great deal to make sure I'm making wise financial decisions. So, when I refinanced my house I did exactly that. I pulled up my big girl panties, faced my fear, and set out to determine what I could do to make the decision and the process feel safer and more manageable. I looked online for the current rates and options for refinancing, called my current lender, and checked with people I knew to see if any of them had any advice or recommendations to share. Bit by bit I became a little more comfortable and less stressed by the prospect of refinancing. In the middle of my searching and researching a friend recommended a mortgage broker who worked with people in my exact situation. What? There were others like me who hated dealing with money? I was thrilled with this news. I called the broker who slowly, carefully, and very kindly walked me through the process. She answered every single question I had and explained things in a way I could understand until I felt comfortable. She calmed my every fear so that I was comfortable and felt safe signing on the dotted line.

If you go to that much work and trouble to calm your fears about something like money, it makes sense to do the same when you are worried about traveling on your own. I can tell you that I will be much less fearful in dealing with mortgage issues the next time than I was the first time. The more you know the "monster" you fear, the less scary he seems. He becomes more familiar and you become better equipped to push him aside and walk bravely onward. Solo travel is no different.

Of course, traveling alone is scary at first. You've never done it before. It's new. You've heard or read about all of the things that can happen, you watch the news and see how horrible things are in some places of the world, and you often doubt your own ability to deal with difficulties that might arise. It's just like when a woman gets pregnant. Suddenly everyone feels a burning need to tell her every pregnancy horror story they've ever heard. Thankfully, most women know not to believe everything they hear and shake their heads at the person who is doing their best, perhaps not intentionally, to scare the heck out of them. Many times, the same thing happens when you share that you are going on a trip alone. The scary stories of vacations gone wrong come flying in from every direction. Instead of listening and letting the monster take over, let's explore some ways for you to fight the monster and march onward toward that journey to your Dream Destination.

In her book, *The Gratitude Connection*, author Amy Collette speaks about fear as a guard dog. She explains that when faced with a new or unfamiliar situation, it is like a stranger knocking at your door. Fear, your good guard dog, runs to the door and barks to warn and protect you. It is your job as the owner to acknowledge the barking and realize that it is merely an early warning system. Fear's "job" is to bark like crazy to warn and protect us. When we are thinking about traveling alone we think that we need to act or decide solely on that fear. But, just as we would never let the dog make the

• Journey for One •

decision of who we should let in our home, neither should we let fear make the decision about what, where, or how we travel. We need to open the door to the fear, see what or who is out there, and then assess for ourselves whether or not the dog had good reason to bark. When you think about your fear as being like a watchdog, it becomes much easier to put yourself in a more powerful position and look at what is best for you, not letting the fear take charge.

Great. Fear is a dog, you say? What do I do to quiet the dog? The first step is to identify the fear. Expose it to the sunlight. Explore what it is, why it is there, and what you need to know so you can decide whether or not to listen and pay attention. In the Pack Your Bag section of this chapter you identified your fears, you described them and explained why you think they are there. Now it's time to explore what you need to know to make decisions based upon fact, not fear.

Let's take a look at a familiar travel scenario. Istanbul, Turkey is one of the biggest, busiest, and most exciting cities in the world. It is the only major city in the world located in two different continents, Europe and Asia. It is located between the Black Sea and the Mediterranean Sea. Elena wants to go to Istanbul on a solo vacation but is scared that it will be dangerous. People start telling her that dangerous things happen there, women are treated poorly, and that women alone can't go anywhere in Istanbul. She gets so scared that she gives up her dream and won't even consider going to Istanbul anymore.

Here is what Elena could have done instead. She could have written down the fearful things that people said or things that she was scared of and then done some research on her own to see if the fear stories were accurate. Once she verifies whether or not her information and fears are true, she can then make an informed decision about her travel plans. It is amazing how often, when we research and check out the things we have heard, they many times turn out to be false or at least exaggerated.

If Elena had written down her fears, researched the information, and then made her decision based upon fact, here's what she would most likely have found.

Question	Facts	Source (all sources in example are fictional)
Is Istanbul dangerous?	No more than any other city. Most terrorist activity is targeted at political sites and at demonstrations, not at tourist destinations.	XYZ Report on World Safety, Volume 1
Are women safe in Istanbul?	Istanbul is one of the safest big cities in Europe for women to travel alone if you stay in the touristic areas, and avoid alleys and empty neighborhoods.	Travel Bug Magazine, October 7
Does Istanbul have a high crime rate?	No. Reports show moderate crime rate. Actually, it is lower than New York City and Paris.	The Bugler Newspaper, Special Edition, 2019

Based upon her new information and looking at many websites, blogs, and social media posts, Elena makes a decision to proceed, with open eyes and reasonable care, to plan her trip to Istanbul.

People who are considering making their first solo journey often have many of the same fears. These fears have come up in one form or another in almost every conversation I have ever had with first-time solo travelers. Look over this list of many of the common fears first-time solo travelers have and I'm sure you will find that your fears are very similar to everyone else's the first time they traveled alone.

- **Safety:** This includes accidents, crime, being alone, talking to strangers, getting stranded or lost, disasters, terrorism, and health.
- **Loneliness**: This includes being shy, embarrassed, homesick, guilt, eating alone, no back up for assistance, and not speaking the language.
- **Creature Comforts**: This includes boredom, logistics, being physically uncomfortable, unfamiliar food, flying, weather, beds, and accommodations.

- **Logistics:** This includes finances, age (I'm too old/young), not speaking the language, finding accommodations, transportation, and getting around.

Your fears are real. The information on which you are basing those fears may, or may not, be real. Before you make a decision, do your best to determine how true and accurate your fears are.

OK, so what if your fears are things that are a little less concrete than eating alone and you're not sure how to research them? What can you do? Here are some examples of those types of fears.
I'm afraid:

- I won't be able to communicate. I don't speak their language.
- I will be lonely.
- I'll get lost in the city and won't be able to find my way to my hotel.
- Because I don't know what to do there.
- Because I don't know how to get to and from the airport.
- Because I don't know a safe place to stay.
- I won't have enough money.

Below are some general guidelines that I have created for myself over the years that have served me well. While they won't solve every problem and take away every fear, they will go a long way in helping you relax enough to calm the fears and carry on planning your solo adventure.

Quiet Your Fears – Tips for Being Proactive

For every fear you have, there are tried and true methods for mitigating that fear. I can't say it will make the fear completely disappear, but it certainly gives you enough insight and strategies to help you walk through the fear and come out on the other side much more informed, aware, and comfortable. Below are my suggestions for dealing with the fear you feel and creating a new mindset that will help you move forward on your solo adventure plans.

Get It Together Guidelines
1. **Start Small**
   Don't make your first solo destination something exotic and complicated. Start with going to another city by driving there. Maybe choose a location that requires a flight if you like to fly, but stay relatively close to home. Keep it simple. The purpose of your first solo adventure is to try it out, to have the experience, not necessarily to focus on the location. That's much easier to do than being far away in some strange location and being totally out of your comfort zone.

2. **Plan Ahead**
   Being successful requires planning. It's like baking a cake and not checking to see if you have all the ingredients in the house, whether or not the oven works, and if you have the time to complete the project. All it takes is a little planning. Do your research about the location. Address your fears early on before leaving home. It's also wise to plan well for the first couple of days so you know what is happening and feel more in control. Free-wheeling, unscheduled traveling is great but I don't recommend it for first timers.

3. **Be Brave**
   One of the greatest gifts of solo traveling is to realize how truly capable you are. We don't usually give ourselves enough credit for being smart, resourceful, and competent. Use those wonderful traits to your advantage. If you want to solo travel, it is important for you to step out of your comfort zone and just go for it. It's like the old saying, "The definition of insanity is doing the same thing over and over again and expecting different results." Traveling on your own is different so you are going to have to think and act differently than you have before. Be strong. Expect different results. You can do it.

4. **Be Positive**
   The American President, Abraham Lincoln, said "Most folks are about as happy as they make up their minds to be." He believed that if you are looking for the bad, the negative, and the unhappy, that's exactly what you are going to find. On the other hand, if you think positive

thoughts, use positive self-talk, and look for the little joys in life, then you are more than likely going to be happy with the world you find. Try to see the good parts of a plan gone wrong. Once, on a road trip in Germany, my mother and I got lost while driving and ended up following a very long, slow line of cars. After a while we were finally turning ... into an open field filled with parked cars and bicyclists. We had unknowingly driven right into the German National Cycling Finals. There was no way out, they told us. It was a one lane road so we would have to wait until the race was over to get back on the highway. Defeated, frustrated, and embarrassed we decided there was nothing to do but settle in and enjoy the view. Before long the other observers started little conversations with us and not long after that we were invited for wine and snacks and spent a lovely afternoon learning German words, cheering for their family members, and having one of the greatest experiences of my life. Had we not been positive, but instead sat in the car and grumbled, we would have missed one of the loveliest cultural experiences ever. It wasn't a disaster after all, just a change of itinerary.

5. **Embrace the Differences**

    One of the reasons people travel is to see things that are different, new, and interesting. I was astounded when I was living in Vienna, Austria how many tourists would come to the city for a visit. They wanted to see Austria, eat Austrian food, and see the amazing history. Yet some just couldn't stop complaining because "They don't speak English here," "Why don't they have food like we have at home?" and my personal favorite, "Why can't they just do it like we do?" We travel to see the world from a new perspective. If you are struggling with fears about things being different than at home, then maybe you need to take a look at what it is you want from travel. I had a wonderful tour guide in Europe once who, whenever anyone complained about things not being what they were used to, would say, "Not bad, just different." What a wise man.

Let's take a look, in closing, at the two definitions below. Fear and without fear or favor are two ways of approaching decision making.

14: Fear

Fear: /fir/, noun: an unpleasant emotion caused by the belief that someone or something is dangerous, likely to cause you pain, or a threat.

Without fear or favor: impartiality, "make all of your decisions without fear or favor."

Making a decision that has an element of fear to it for you is best made without fear or favor: that it is made in an impartial manner, only after you have checked out the facts, either proving or disproving your fears. Know and face the monster. Then proceed bravely with confidence and knowledge.

Adventure

Now, as a way to put your new-found knowledge, skill, and desire into action, I want you to do a little exploration. This exploration will be both inside of yourself and by using the internet or other resources.

Take a look at the fears that you labeled as threes in the Pack Your Bag section of this chapter. Those are the things you feel are the most frightening, the things that scare you the most about traveling alone to your Dream Destination. In order to face those fears, acknowledge them, and then make an informed decision, I want you to continue by researching the fears you addressed. Pull out each of the fears that you labeled a three on the chart. Then turn the fear into a question that you will then research. Make notes about the information you find even if the information may be different when coming from different sources. Use the chart about Istanbul in this chapter's Learning the Way section as a guide. I have used Elena's information from above as a sample here.

• Journey for One •

Dream Destination:

Question	Facts	Source
Is Istanbul dangerous?	No more than any other city. Most terrorist activity is targeted at political sites and at demonstrations, not at tourist destinations.	XYZ Report on World Safety, Volume 1
Are women safe in Istanbul?	Istanbul is one of the safest big cities in Europe for women to travel alone if you stay in the touristic areas, and avoid alleys and empty neighborhoods.	Travel Bug Magazine, October 7
Does Istanbul have a high crime rate?	No. Reports show moderate crime rate. Actually, it is lower than New York City and Paris.	The Bugler Newspaper, Special Edition, 2019

• 14: Fear •

Looking in the Rearview Mirror

How surprised were you by the results of your research? Have a serious and honest conversation with yourself. Give these questions some thought. Jot down your answers so you can refer back to them as you travel and learn more. Remembering your early thoughts as you try new skills is important to your growth as a traveler.

Research Reflection

How did you do?	
What did you notice?	
How did you feel?	
What would you do differently or more of next time?	

Celebration

Congratulations! You are one giant step closer to being a much more courageous and skilled solo traveler.

I want you to think about somewhere in your own community or area where you have never been. Think about a place that you know you would enjoy. It may be hiking, having lunch, a museum, lecture, or whatever you like. Please make it something that allows you to be able to observe other people. Attending the opening of a new coffee shop qualifies for this activity and so does going to a museum, art show, sporting event, or attending a festival or celebration.

Now, identify your fears, and think about or research the truth and/or reality of those fears. Once you have addressed those fears, stop the barking guard dog and walk bravely and confidently out your door and enjoy your latest Journey for One.

I want you to treat yourself to this activity alone. I want you to start the journey at home, in your own mind, before you ever set foot out of the door. Take a moment to think of where you will go and set a couple of goals for yourself. Examples of those goals are:

- Stay one hour.
- Make contact with someone new and have a short conversation.
- Initiate a conversation by asking someone a question.
- Observe the number of people who are also there solo.
- Practice calming breathing.
- Make a list of things that seem new or interesting to you.

15

READY, SET, GO

Kid, you'll move mountains! Today is your day!
Your mountain is waiting. So, get on your way.

—Dr. Seuss

Travel Tale

This past spring, I had the joy of attending two graduations. One was for my niece who was graduating from college and the other was for a young friend who was graduating from high school. As I sat through the usual boring graduation speeches, I smiled, thinking of my own graduations in the past and what a mix of emotions I had experienced each time. I was, at the same time, thrilled, hopeful, exhausted, scared, full of confidence, and full of expectations for a bright and beautiful future ahead. I knew that my two graduates were feeling very much the same from chats I'd had with them. I smiled and felt myself giggle a bit while enduring the graduation speeches at how your upcoming completion of this Journey for One journey would be similar. You will most likely experience many of the same emotions as a graduate does when you book that first solo trip and prepare for this next big step in your adventure toward traveling alone. You have been on solid ground and you have now gathered up your courage, knowledge, and desire and are ready to take that next first leap across the stream toward your first solo journey.

I remember sitting on the airplane those many years ago heading to that first conference in St. Louis. I had stayed alone in a hotel. I had eaten by myself in a restaurant. I had explored a new town on my own and, most importantly, I had even started to believe I might like it. As the plane left the ground and I watched us climb into the clouds, I felt new, different. Somehow, I knew this was the start of something important. I had no idea how or why it would affect me but I was willing to believe in myself and my ability to be successful. I knew that if I didn't give this solo travel thing my very best effort that I would regret it forever.

As that plane glided effortlessly through the air, I remembered a story about my great-grandmother, Kathryn, a farmer's wife in a teeny tiny farm town in Kansas. She and my great-grandfather had cared for and lived on their small wheat farm their entire married life. Great-grandma was now in her 90s and in a nursing home.

Because it was such a small town, every one of the residents of the nursing home had known each other most of their lives. Great-grandma's roommate, Mrs. Jones, happened to be the wife of the bank president and was probably the richest woman in town. One day as my mom was visiting, her grandmother turned to her and said, "You know, I'm richer than Mrs. Jones." My mother became instantly concerned about Great-grandma's mental abilities. "Oh, I know she has more money in the bank than I do, but my scrapbooks are thicker."

This story has been one of my favorites throughout my life. I found it explained my thoughts about why I travel, why I value the people in my life, and why I say yes to almost every adventure and experience I encounter. I even made a framed print that says, "She with the thickest scrapbook wins." I want to go out a winner. I want to leave behind a house full of very thick scrapbooks. Sitting on that plane, headed for my first solo trip to St. Louis, I was creating an important new page in my life scrapbook. I think Great-grandma would be proud.

Pack Your Bag

The end of our journey together through this book together is coming to a close. I hope that this journey has given you insight, ideas, tools, and the courage to now strike out on your own. If there is one word that I believe makes the biggest difference between success and failure as I travel on my own, it is *prepare*. Prepare for the best outcome. Prepare for every possibility. Prepare so you get what you need. Prepare so that you aren't disappointed. And last but not least, prepare to be changed forever by your travel experiences.

It is now time for you to take the lead in creating your solo travel life. You have learned some new skills and gained new ideas, but one book cannot be all things to all people. It's time for you to prepare on your own.

In the space below, collect your thoughts about what lies ahead. What have you decided? Are you ready to head out on your own or are you not yet sure that solo travel is for you? Either way, you know the answer and it is the right one for you. Take this time to put in writing what you need now. What are the next steps that you want to take to prepare for your travel future? How can you find and obtain what you need in order to have a successful first solo trip? This is important. Once you set this book down it is easy to get busy with other things and let this new information slide to the back of your mind. You have worked hard and you deserve to remember the great work you have done. So, what now?

_____

_____

_____

_____

_____

_____

_____

_____

_____

_____

Learning the Way

Oh, my goodness! How time flies when you're having fun. We are now on the last lesson of *Journey for One*. I hope that you have found what you needed to help you begin your solo travels.

Just as we often do with those we are sending off on a new adventure, I want to leave you with a bit of sage advice to take with you as you move forward: little nuggets of wisdom to light your path.

- Your first solo journey is best taken somewhere not too far from home. Plan a little weekend away to someplace you enjoy and perhaps know a bit. Being at a location fairly close to home and familiar gives you the option of going home early if you find that you aren't as ready as you thought you were to be going it alone. There's no shame in this. It's just a heck of a lot easier and cheaper the closer you are to home the first time.
- Be willing to flail. Remember, having the freedom to flail means that you can, and probably will, make some mistakes, enjoy some parts and not others, and you may find that what you thought you wanted or needed was different once you are there. It's perfectly OK to feel uncomfortable and get frustrated. Just remember to breathe. Give yourself the gift of time and patience. You can do this!
- Start out with a positive attitude. Remember Henry Ford's saying about "Whether you believe you can do a thing or not, you are right"? Well, it's never been truer than when you are trying out something new and a bit scary. Keep telling yourself you can do this. Don't sabotage yourself with negative thoughts and self-talk.
- Keep a journal of your first solo outing. I didn't and truly wish I had. It would be fun to look back and see what I did, what I thought, and how I navigated through my first solo trip. It will be fun for you to look at your notes after a few solo travels and celebrate your amazing growth.
- Utilize the KISS principle: Keep It Simple, Silly. The more things you try to do, the more possibilities there are for mix ups. Make your first solo trip a weekend outing. Plan simple and enjoyable activities that you love. Maybe get a massage or even do a whole spa day. Go for a hike. Watch a basketball game. Do some shopping. The simpler you

keep your itinerary, the more relaxed you will be, and the more you will enjoy your journey.
- Take at least one photo of yourself. I know, I hate looking at pictures of myself too, but it will be important after you return home to have the documentation for yourself and the ability to look back and have proof positive that you really did travel alone.
- I encourage you to try to enjoy at least one meal out at a restaurant while you are on your first voyage. It is easy when we are alone to do the whole drive-through, fast-food thing but we don't get better and grow unless we push ourselves a bit. I'm not talking about a five-star dining experience here unless that's what you enjoy. A simple cafe or pancake house is perfectly fine. (A little hint: people find that eating breakfast or lunch out alone is a bit more comfortable than dinner. I'm not sure why, but it might be worth a try.)
- Take this book with you when you make your first trip. It will help if you begin to struggle a bit to have the information and your reflections at hand to help you calm yourself and remember what you have learned.
- Last but not least: have fun. Find reasons to smile. Laugh as often as you can, even if it's at yourself. You don't see the joy in the world around you unless you look for it.

Adventure

One of the ways to discover what you want and the options you have in life is to look at the examples of others. As you have moved through the activities in this book you have had the opportunity to hear from me, friends, and other travelers about their experiences and advice. There is another option, which I admit isn't completely factual in its information, but which gives you ideas and opens your mind to the possibilities and benefits of solo travel.

Your assignment for this adventure is to take a look at the list of movies below. They have been suggested by various sources as films that portray solo travel in interesting ways. I have personally seen each of these films and can honestly say that they offer a great deal in terms of insight, possibilities, and the pros and cons of traveling alone. While some are Hollywood versions of traveling on your own and have romanticized the adventures, the basic message of the power of solo travel remains. So, grab the popcorn, put your feet up, and watch at least one, hopefully more, of these films and pay attention to the thoughts that come to mind about possible implications or ideas for your own journeys.

Movie Title	Year	Details
Into the Wild	2007	Director: Sean Penn After graduating from college, a young man gives away all of his possessions and hitchhikes to Alaska to live in the wilderness. It has a sad ending but is a wonderful film full of interesting characters.
Eat, Pray, Love	2010	Starring: Julia Roberts After a painful divorce the main character sets out on an around-the-world journey to find herself.
Tracks	2013	This is an award-winning Australian film about a woman's journey across the West Australian desert with four camels and her dog.
Wild	2014	Starring: Reece Witherspoon A powerful film about a woman who makes a 1,100-mile trek by herself to deal with a personal tragedy.
The Motorcycle Diaries	2004	This is an Oscar-winning film that tells the story of Che Guevara, an Argentinian revolutionary during the Cuban Revolution, and the motorcycle road trip that led him to his passion and purpose in life.
The Secret Life of Walter Mitty	2013	Directed by and Starring: Ben Stiller. When the main character's job is threatened, he takes off on a huge journey that becomes an unbelievable adventure.
One Week	2008	A Canadian film about a cross-country motorcycle trip made by Ben Tyler. It chronicles his quest to find meaning in his life.

How Stella Got Her Groove Back	1998	A bit silly, but a fun movie about Stella, who goes to Jamaica to explore her life and find a balance between her work, motherhood, and her love life.
Kumiko, The Treasure Hunter	2014	This movie depicts a Japanese woman who believes that a video of the movie *Fargo* is the key to a hidden treasure in Minnesota. She takes off alone from Japan and travels to Minnesota to find the treasure.
The Way	2010	Starring Emilio Estevez and Martin Sheen. After his son dies while hiking El Camino de Santiago in Spain, a father goes overseas and decides to hike the trail himself in search of peace and meaning in his son's death. Very moving.
Under the Tuscan Sun	2003	Starring: Diane Lane A writer leaves the United States and impulsively buys a villa in Tuscany, Italy. The experience changes her life.

Looking in the Rearview Mirror

Here's your chance. Author Pattie Mallette says, "... everyone's a critic," and I'm giving you the chance to do just that. Pick a couple of the films that you watched and write up your thoughts, especially those that showed you something important to you about your own future travel. Use this opportunity to have fun imagining what fun or opportunities you might have that would make a good movie of your travels.

Film # 1 Title:

Important thoughts and ideas from the film for me	
Fun ideas and opportunities I got from the film	

Film #2 Title:

Important thoughts and ideas from the film for me	
Fun ideas and opportunities I got from the film	

Celebration

Cue the "Pomp and Circumstance" music for graduation! This is it. You took on the challenge of gaining the courage and skills to travel solo and saw it through to the end. Regardless of your final decision about traveling alone, you have succeeded. You have given a great deal of yourself and your time to this self-exploration and deserve a lot of credit for having made the journey.

You are now in a place where you can make an informed and personal decision. You can take all the time you need to decide if taking off on a solo adventure is something you want to do. I have tremendous respect for you and celebrate whatever travel lies ahead for you.

No matter whether you decide to travel solo or if you decide you'd rather not be on your own, I want you to keep traveling. Keep trying. Keep taking those important first steps. In the end, it is not the destination that matters, it is the journey.

I offer you this final, closing thought. It is a quote from Andrew Zimmern:

"Please be a traveler, not a tourist. Try new things, meet new people and look beyond what's right in front of you. Those are the keys to understanding this amazing world we live in."

Happy and safe travels, my friend. May your days be filled with joy, wonder, and unforgettable experiences.

Thank you for being a part of *my* journey.

ABOUT THE AUTHOR

Jodie Hopkins started traveling alone at the age of 5 when she ran away from home with cookies in a bandana tied to a stick. Since then she has traveled as a student, volunteer, teacher, trainer, and typical tourist to 33 countries. She has visited six continents and lived on her own in Austria for 11 years. Jodie speaks English, Spanish, and German. Her love of travel has led her to volunteer numerous times as a translator for medical mission trips, teach overseas in an international school, and play traveler all over the world. She says she wouldn't trade one minute of her travel experiences even if she could.

While she now visits locations as exotic as India, Portugal, South Africa, and Amazonian Ecuador, New Zealand and Costa Rica are tied as her favorite locations so far. Who knows where she might be off to next?

When she is not traveling or writing, Jodie lives in Colorado enjoying learning new things, hanging out with friends and family, and absolutely loving life. She is a devoted daughter and sister, and knows she is the luckiest aunt in the whole world.

Jodie Hopkins can be reached by email at journeyforonego@gmail.com.

ACKNOWLEDGMENTS

This book was a work of love. It represents both my love of travel and my love of teaching. No important work is ever done in isolation. I received tremendous love and support from the many people and could not have seen it through without those in my life who have brightened my path and enriched my journey.

I want to express tremendous gratitude to ...

Delene for being my cheederleader and champion. (The spelling is intentional.)

Shely for always believing in me and getting me to believe in myself.

Brandon for always keeping me supplied with smiles, laughs, and love. Grovesnor hot dogs forever!

Shawn for being die beste Nichte der ganze Welt. Ich liebe dich!

My mom for teaching me to make it fun and be self-sufficient, and for our millions of travel miles together.

My Vienna Family for giving me a home, family, and unbelievable memories of love.
  The Radevs: Dima, Radosvet, Rada, Sveta, Bobby, and Rosie
  The Goldens: Brian, Tanja, Romy, and Liah
  Hanrich Claassen and Kent Merdes
  Michael Niederer and Andreas Wessely
  Mario Zrno
  Kelly and Yves Underwood
  Mark Kolinski and Ellen Levenhagen
  Ginny Sampson and Jeff Audette
  Gordon Augustine

Karin Q-D for your insight, compassion, and undying encouragement. You made all the difference.

The kids that I had the joy of teaching for inspiring me and making me want to be a better me.

The Positively Powered Authors Group for your encouragement and inspiration.

Grandma Wodark for making her journey so I could grow to cherish mine.

Lucy for all the early morning hours cuddled at my feet while I wrote. It made all the difference.

Amy for whom I am incredibly grateful. You found and nurtured my passion and gave me the dream of my name on a book cover. You believed in me, supported me, and made me believe it would happen. You are a priceless treasure.

...and with loving gratitude to Dale for being a part of our family and for keeping me company while I wrote.

Thank you all for being a very special part of my journey.

RESOURCES

Articles:
    Brilliant Living HQ. "6 reasons why you should celebrate success." https://www.brilliantlivinghq.com/6-reasons-why-you-should-celebrate-success

    Sahaj, Kohli. "22 Pieces of Advice For First-Time Solo Travelers." *Huffington Post*. Updated July 22, 2016. https://www.huffpost.com/entry/22-pieces-of-advice-for-first-time-solo-travelers_n_5787ac2de4b0867123dff597

Books:
    Lonely Planet. *The Solo Travel Handbook.* Lonely Planet, 2018.
    Waugh, Janice Leith. *The Solo Traveler's Handbook*. 2nd ed. Full Flight Press, 2012.

Websites/Social Media:
    The Solo Female Traveler Network, Facebook
    Over 60 Solo Women Travelers, Facebook
    Centers for Disease Control, Travelers Health Section, https://wwwnc.cdc.gov/travel
    U.S. Department of State, Travelers Section, https://www.state.gov/travelers

Made in the USA
Monee, IL
17 July 2021